COMMON CORE
STANDARDS

for | Middle School
Mathematics

A QUICK-START GUIDE

Edited by John Kendall

COMMON CORE STANDARDS

for Middle School Mathematics

Amitra Schwols

Kathleen Dempsey

ASCD

Alexandria, Virginia USA

McREL

Mid-continent Research for Education and Learning
Denver, Colorado USA

1703 N. Beauregard St. • Alexandria, VA 22311-1714 USA
Phone: 800-933-2723 or 703-578-9600 • Fax: 703-575-5400
Website: www.ascd.org • E-mail: member@ascd.org
Author guidelines: www.ascd.org/write

MREL

Mid-continent Research for Education and Learning
4601 DTC Boulevard, Suite 500
Denver, CO 80237 USA
Phone: 303-337-0990 • Fax: 303-337-3005
Website: www.mcrel.org • E-mail: info@mcrel.org

All web links in this book are correct as of the publication date below but may have become inactive or otherwise modified since that time. If you notice a deactivated or changed link, please e-mail books@ascd.org with the words "Link Update" in the subject line. In your message, please specify the web link, the book title, and the page number on which the link appears.

PAPERBACK ISBN: 978-1-4166-1464-7   ASCD product #113013   n1/13
Also available as an e-book (see Books in Print for the ISBNs).

Quantity discounts: 10–49 copies, 10%; 50+ copies, 15%; for 1,000 or more copies, call 800-933-2723, ext. 5634, or 703-575-5634. For desk copies: www.ascd.org/deskcopy.

Library of Congress Cataloging-in-Publication Data
Schwols, Amitra.
  Common core standards for middle school mathematics : a quick-start guide / Amitra Schwols, Kathleen Dempsey ; edited by John Kendall.
      p. cm.
  Includes bibliographical references.
  ISBN 978-1-4166-1464-7 (pbk.)
  1. Mathematics—Study and teaching (Middle school—Standards. I. Dempsey, Kathleen, 1952- II. Kendall, John S., 1949- III. Title.
  QA11.2.S394 2013
  510.71′2—dc23
                              2012043236

_____
22 21 20 19 18 17 16 15 14 13 1 2 3 4 5 6 7 8 9 10 11 12

COMMON CORE STANDARDS

for | Middle School Mathematics

Acknowledgments

We would like to acknowledge Kirsten Miller and John Kendall for their crucial role in making our thoughts much more readable; Greg Gallagher and the North Dakota Curriculum Initiative committee, who provided us with valuable insights into the challenges facing teachers as they begin to work with the Common Core standards; Ceri Dean for her step-by-step guide to lesson planning; Amber Evenson for her collaboration and content expertise in developing the lessons; our McREL colleagues, who provided an analytical ear as we discussed the work; and our families, for supporting us as we worked on this project.

Introduction

In July 2009, nearly all state school superintendents and the nation's governors joined in an effort to identify a common set of standards in mathematics and English language arts (ELA), with the goal of providing a clear, shared set of expectations that would prepare students for success in both college and career. The Common Core State Standards Initiative (CCSSI) brought together researchers, academics, teachers, and others who routed multiple drafts of the standards to representatives including curriculum directors, content specialists, and technical advisors from all participating state departments of education. By spring 2010, drafts were submitted for comment to the national subject-area organizations and posted for public comment. In June 2010, the final versions were posted to a dedicated website: www.corestandards.org. (A minor update of the standards was posted in October 2010.)

At press time, 45 states, as well as Washington, D.C., and two territories, have adopted the Common Core State Standards (CCSS) for mathematics. (Minnesota has adopted the ELA standards but not the mathematics standards. Texas, Alaska, Virginia, and Nebraska have indicated that they do not plan to adopt either set, although both Virginia and Nebraska have aligned the Common Core standards with their existing standards.)

Adoption of the standards is, of course, voluntary for states and does not include a commitment to any other programs or policies. However, states that have adopted these standards will be eligible to join one of two federally funded assessment consortia that are currently tasked with developing assessments for the Common Core—the Smarter Balanced Assessment Consortium (SBAC) or the Partnership for Assessment of Readiness for College and Careers (PARCC). Sharing assessments across states promises financial relief from notoriously expensive state assessments. In addition, federal programs such as Race to the Top have required that applicants demonstrate that they have joined with other states in adopting a common set of standards and an assessment program. Although states may form new consortia, many either have opted to join or are considering joining SBAC or PARCC.

Sharing a set of standards across states offers other advantages. For example, teachers' well-designed lesson plans targeting Common Core standards will be immediately useful to a large number of colleagues. The shared language of standards should also provide teachers with more opportunities to participate in very specific discussions about content, a process that has been hampered somewhat by the variety of ways states have described virtually the same content.

For a lengthier discussion of the Common Core standards, including their link to previous standards-based education efforts and the benefits and challenges the Common Core presents, see *Understanding Common Core State Standards* (Kendall, 2011), the first booklet in this series. We also encourage readers to explore numerous resources available at corestandards.org, especially the standards document itself (CCSSI, 2010c) and the guidelines for adapting standards instruction for English language learners (CCSSI, 2010a) and students with disabilities (CCSSI, 2010b).

About This Guide

This guide is part of a series intended to further the discussion and understanding of Common Core standards on a subject-specific and grade-level basis and to provide immediate guidance to teachers who must either adapt existing lessons and activities to incorporate the Common Core or develop new lessons to teach concepts not addressed in their previous state standards.

After an overview of the general structure of the Common Core standards for middle school mathematics, we consider each domain in turn, grade level by grade level, to examine how the standards it contains build upon and extend the skills students have acquired in earlier grades. We also explore the links between domains and highlight the mathematical practice standards with the strongest connections to each domain. Next, we focus on practical lesson planning with the Common Core, looking at a process for creating standards-based lessons that make the best use of the effective instructional strategies explored in *Classroom Instruction That Works, 2nd edition* (Dean, Hubbell, Pitler, & Stone, 2012). The guide concludes with an illustration of this process's outcome: three sample lessons that address Common Core standards identified as representing notable changes to middle school mathematics teachers' current practice.

CHAPTER 1

About the Common Core Mathematics Standards for Middle School

The Common Core mathematics standards are organized into two sets: the Standards for Mathematical Content, designed to cross traditional course boundaries and cover all the conceptual mathematical understandings students need to develop from kindergarten through 12th grade, and the Standards for Mathematical Practice, which highlight the kinds of expertise that students must develop and use throughout this same grade span.

As we will show in this guide, the Common Core standards differ in many ways from most existing state standards documents, providing a greater level of detail about concepts, thought processes, and approaches. This level of detail often leads to much longer, more involved standards, some of which are up to a paragraph in length. Some of the standards detail conceptual methods of teaching and learning skills and concepts (e.g., applying the properties of operations to generate equivalent expressions, understanding that solving an equation or inequality is a process of answering a question). This is in stark contrast to many prior sets of state standards, which were far less explicit and typically used a single sentence to describe the skills and knowledge required of students. Another example of this detailed focus on the mental processes required in understanding mathematical concepts is found in the set of Standards for Mathematical Practice, which receives the same level of emphasis as the Standards for Mathematical Content.

In this chapter, we will walk you through the standards' structure, provide an overview of how the middle school mathematics standards fit together, and offer some guidance on what to focus on as you begin your implementation efforts.

The Standards for Mathematical Content

At the middle school level, the Standards for Mathematical Content are organized first by grade, then by domain, and finally by cluster. Each grade level's set of standards is introduced with a one- or two-page introduction, which consists of two parts—a summary of the three to four critical areas (topics) for each grade, and an in-depth narrative description of those critical areas. Figure 1.1 provides a brief, grade-by-grade summary of the critical areas for middle school.

One important aspect of the introductions is that their narrative descriptions of the critical areas also provide insight into the meanings or limitations of the standards that are not immediately apparent to someone just reading the standards alone. For example, the first standard in the 6th grade Ratios and Proportions domain asks students to "understand the concept of a ratio"—quite a broad aim. The introduction to the 6th grade standards (CCSSI, 2010c, p. 39) provides vital clarification, articulating in greater detail which aspects of the concepts of ratio a 6th grade student needs to understand, connecting ratios, rates, and an ability to reason with multiplication and division. Although it can be tempting to skim introductory text, teachers and administrators should take the time to review the standards' grade-level introductions thoroughly to ensure that they get all the information about the standards that is available.

Figure 1.1	**Critical Areas Within the Middle School Mathematics Domains by Grade Level**		
Domain Name	**Grade 6**	**Grade 7**	**Grade 8**
Ratios and Proportional Relationships	Connecting ratio and rate to whole number multiplication and division and using concepts of ratio and rate to solve problems	Developing understanding of and applying proportional relationships	*Domain not addressed at this grade level*
The Number System	Completing understanding of division of fractions and extending the notion of number to the system of rational numbers, which includes negative numbers	Developing understanding of operations with rational numbers and working with expressions and linear equations	*No critical areas identified for this domain at this grade level*
Expressions and Equations	Writing, interpreting, and using expressions and equations		Formulating and reasoning about expressions and equations, including solving linear equations and systems of linear equations
Functions	*Domain not addressed at these grade levels*		Grasping the concept of a function and using functions to describe quantitative relationships

(continued)

Figure 1.1 | **Critical Areas Within the Middle School Mathematics Domains by Grade Level** *(continued)*

Domain Name	Grade 6	Grade 7	Grade 8
Geometry	*No critical areas identified for this domain at this grade level*	Solving problems involving scale drawings and informal geometric constructions, and working with two- and three-dimensional shapes to solve problems involving area, surface area, and volume	Analyzing two- and three-dimensional space and figures using distance, angle, similarity, and congruence, and understanding and applying the Pythagorean Theorem
Statistics and Probability	Developing understanding of statistical thinking	Drawing inferences about populations based on samples	*No critical areas identified for this domain at this grade level*

Note: Content in this table was adapted from the descriptions in the standards' grade-level introductions.

After each grade-level introduction, the standards are organized hierarchically, as follows:

• *Domain:* Expressed in one or two words, a domain articulates big ideas that connect standards and topics. The middle school standards are categorized into six domains: Ratios and Proportional Relationships (PR), the Number System (NS), Expressions and Equations (EE), Functions (F), Geometry (G), and Statistics and Probability (SP).

• *Cluster:* A cluster captures several ideas that, taken with all the other clusters within a domain at a grade level, summarize the important aspects of mathematics students will encounter during the year. For example, the Expressions and Equations domain at the 8th grade level has three clusters. The first ("Cluster A") is "Work with radicals and integer exponents." The second

("Cluster B") is "Understand the connections between proportional relationships, lines, and linear equations." The third ("Cluster C") is "Analyze and solve linear equations and pairs of simultaneous linear equations." The content addressed in different domains and clusters may be closely related, reflecting the standards writers' emphasis on the interconnections throughout mathematics.

• *Standard:* A standard is a specific description of what students should understand and be able to do. Each standard is numbered. It may be one sentence or several sentences long, and it sometimes includes lettered components. Typically, there are one to several standards within every cluster, although the standards themselves are numbered sequentially within the domain. For example, the first of two standards in Cluster B of the Expressions and Equations domain at the 8th grade level, which is Standard 5 within the domain overall, is "Graph proportional relationships, interpreting the unit rate as the slope of the graph. Compare two different proportional relationships represented in different ways."

This guide contains one chapter for each of the six middle school domains, and at the beginning of each chapter, you will find a chart that provides an overview of each grade level's clusters and standards. In those charts, and throughout this guide, we will be referencing the content standards using a slightly abbreviated version of the CCSSI's official identification system, which provides a unique identifier for each standard in the Common Core and can be very useful for school staffs when developing crosswalks, planning lessons, and sharing lesson plans. Under this system, all mathematics content standards begin with the formal prefix "CCSS. Math.Content"; we have dropped this prefix in our references throughout the guide, including the sample lessons. The next piece of the code for standards in grades K–8 is the specific grade level, which is followed by the domain abbreviation, the letter identifying the particular cluster within the domain, and then the specific standard number. For example, "8.G.B.6" is shorthand for Grade 8, Geometry (the domain name), Cluster B (of the domain's three clusters, identified A–C), Standard 6. "8.G.B" is a shorthand way of referring to all the standards within Cluster B of the 8th grade Geometry domain. Note that codes for high school mathematics standards follow the "CCSS.Math.Content" prefix with the letters "HS" and the abbreviation for the conceptual category before continuing with domain, cluster, and standard identifiers. For example, "HSA-REI.A.1" refers to Standard 1 within the first cluster ("Understand solving equations as a process of reasoning and explain the reasoning") of the Reasoning with Equations and Inequalities domain of the high school algebra standards.

Taken as a whole, the Common Core's mathematical content standards at the middle school level identify what students should know and be able to do in order to be prepared for mathematics study at the high school level and, ultimately, to be college and career ready.

The Standards for Mathematical Practice

Emphasis on students' conceptual understanding of mathematics is an aspect of the Common Core standards that sets them apart from many state standards. However, the eight Standards for Mathematical Practice, listed in Figure 1.2, play an important role in ensuring that students are engaged in the actual *use* of mathematics, not just in the acquisition of knowledge about the discipline. Indeed, the table of contents in the standards document gives equal weight to the Standards for Mathematical Practice and to the Standards for Mathematical Content. This dual focus, echoed throughout the standards document's introductory material, has been undertaken to ensure the standards "describe varieties of expertise that mathematics educators at all levels should seek to develop in their students" (CCSSI, 2010c, p. 6).

The writers of the Common Core describe these practice standards in an introduction, explaining that the standards are derived from the process standards of the National Council of Teachers of Mathematics (NCTM, 2000) and the strands of mathematical proficiency found in the National Research Council report *Adding It Up* (2001). A brief description of the meaning of the practice standards is provided in the front of the mathematical standards document (CCSSI, 2010c, pp. 6–8).

Figure 1.2 | **The Standards for Mathematical Practice**

MP1. Make sense of problems and persevere in solving them.

MP2. Reason abstractly and quantitatively.

MP3. Construct viable arguments and critique the reasoning of others.

MP4. Model with mathematics.

MP5. Use appropriate tools strategically.

MP6. Attend to precision.

MP7. Look for and make use of structure.

MP8. Look for and express regularity in repeated reasoning.

In addition to stressing mathematics proficiencies that cross all domains, the mathematical practice standards ensure that students who are focused on skills and processes don't find themselves engaged in rote activities that provide them no deeper sense of how mathematics works as an integrated whole. For example, solving simple equations or inequalities might be seen as nothing more than a series of steps. In the past, standards documents required nothing more than that the process be taught, which meant that often students were shown processes (such as the substitution method) and expected to memorize them. What students were not expressly expected to do was gain a deep understanding of the reasoning and meaning behind the processes. In contrast, the Common Core standards require that students understand that solving equations and inequalities means answering the question "Which values from a specified set make the equation or inequality true?" The Common Core standards ask that students possess this understanding in addition to being able to use substitution (6.EE.B.5). The underlying rationale is that students who are able to articulate this understanding are positioned to gain a deeper understanding of equations and inequalities, which allows them to see the utility of the process over a wider range of problems.

Please note that, as with the content standards, the mathematical practice standards have official identifiers, which we have shortened in this guide's sample lessons. For example, we abbreviate Mathematical Practice 1, officially "CCSS.Math.Practice.MP1," as "MP1."

Implications for Teaching and Learning

A recent survey of more than 13,000 K–12 math teachers and 600 district curriculum directors across 40 states shows that teachers are highly supportive of the Common Core standards. That's the good news. On the other hand, the same survey shows that an overwhelming majority (80%) mistakenly believe that the standards are "pretty much the same" as their former state standards, and only about 25 percent of respondents are willing to stop teaching a topic that they currently teach, even if the Common Core State Standards do not support teaching that topic in their current grade (Schmidt, 2012).

These findings suggest some damaging possible consequences. If teachers don't recognize the Common Core's new emphasis on depth of content understanding, they may not take the steps necessary to narrow the focus of their curriculum so that students will have the time they need to develop that deeper understanding. Furthermore, teachers' unwillingness to stop teaching familiar or favorite content that the standards do not require reinforces the possibility that, while students may be exposed to a wide variety of mathematical concepts, they will not reach the required level of mastery set for concepts that have been identified as critical.

We want to highlight two documents that can provide significant support for teachers' instructional efforts. The first is Appendix A of the standards document (CCSSI, 2010d), which includes information on an accelerated pathway for middle school. This pathway illustrates how to compress the contents of a first-year high school mathematics course (Algebra I or Integrated Pathway: Mathematics I) into middle school. The second useful document is *Progressions for the Common Core State Standards in Mathematics* (Common Core Standards Writing Team, 2011). Still in draft form at the time of this writing, it details some useful strategies for teaching the middle school mathematics standards, and we urge anyone interested in specific strategies and examples to read it.

How to Begin Implementation

As noted, the standards document and its appendix do offer some ideas for how to get started planning instruction and teaching the standards, and here in this guide, we share our own best advice.

Focus on the mathematical practice standards

The Standards for Mathematical Practice are one of the potentially challenging aspects of Common Core implementation. As described on page 6, the mathematical practice standards are found in two places in the standards document: in the document's introduction and in the overview of each grade. The guidance found in the main introduction provides valuable insight into each mathematical practice standard, and we recommend that teachers become extremely familiar with these descriptions and spend some time planning how to incorporate the practices into each course. In the chapters to come, we offer our own ideas about how teachers might integrate the mathematical practice standards with each of the domains in the mathematical content standards.

Focus on critical areas

By sharpening the focus of each grade on three to four critical areas identified by the Common Core writers (as described on pp. 7–8), teachers can develop students' understanding of those concepts to a degree that's deeper than previous standards required or allowed. The outcome is stronger foundational knowledge.

Focus on connections

Remember that the Common Core mathematics standards are designed to be coherent within and across grades. The chapters on the domains clarify how the concepts found in the middle school standards are organized across grades, underscoring that each standard is best understood not as new knowledge but as extensions of ideas presented in previous school years.

The Standards for Mathematical Practice provide further connective tissue between the standards at each grade level and within the various domains, which we highlight throughout this guide. However, it is important to stress that what we present are only a few examples of such connections; we do not mean to suggest that no other connections can or should be made. We encourage teachers to build on the proposals here to strengthen their own practice and enhance their implementation of the Common Core standards.

* * *

Now that we've looked at the overall structure of the Common Core standards for middle school mathematics, we will examine each domain, addressing the specific standards at the various grade levels.

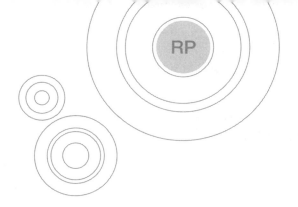

CHAPTER 2

Ratios and Proportional Relationships

The Ratios and Proportional Relationships domain (RP) is found in the mathematics standards for grades 6 and 7 only. In the standards document's introduction to this domain, ratios and proportional relationships are identified as critical content at both grade levels, marking the domain as an area of emphasis for both 6th grade and 7th grade teachers.

The standards in this domain introduce students to the concept of ratios and proportion, focusing first on connecting the concepts of ratio and rate to whole-number multiplication and division, then on using ratio and rate to solve problems, and finally on developing an understanding of proportional relationships and applying that understanding to solve problems. Figure 2.1 shows an overview of the Ratios and Proportional Relationships domain for middle school.

In this chapter and in all the domain-focused chapters that follow (Chapters 3–7), we will first look at how the content of the domain relates to the Standards for Mathematical Practice and then provide an overview of how the content of each cluster relates to the other standards, both within and across grades. This close analysis is intended to clarify the meaning of each standard within the context of the entirety of the Common Core

	Figure 2.1	**The Ratios and Proportional Relationships Domain: Middle School Overview**		

Grade Level	Clusters	Standards
Grade 6	**6.RP.A** Understand ratio concepts and use ratio reasoning to solve problems.	6.RP.A.1, 6.RP.A.2, 6.RP.A.3
Grade 7	**7.RP.A** Analyze proportional relationships and use them to solve real-world and mathematical problems.	7.RP.A.1, 7.RP.A.2, 7.RP.A.3

standards for mathematics and illustrate how students have been prepared for this content through the standards that appear in earlier grades.

Connections to the Standards for Mathematical Practice

The Ratios and Proportional Relationships domain focuses on foundational understanding of ratios and proportions. Students learn what ratios and proportions are, how they're symbolized, and how to interpret them regardless of representation style. As students begin to make sense of ratios and proportions, they are asked to use their new understanding to solve multi-step ratio, percent, and proportion problems.

To facilitate the deeper understanding of ratios and proportions needed to solve these problems, teachers can assign complex tasks that help students see the need to go beyond a superficial approach. These problems should require students to demonstrate a deep, genuine understanding of the targeted mathematical concepts and should allow for multiple solution methods. For example, after students understand the concept of a ratio and can use ratio language, the topic of Standard 1 for 6th grade (6.RP.A.1), a teacher might ask them to figure out how much money a person will spend in a given week on gasoline, given the fuel efficiency of the car, the cost per gallon, and the number of miles that will be traveled. The complexity of this problem can be raised by allowing students to decide between gas stations—one that does not require extra travel but is more expensive or one that is farther away but cheaper. (See NCTM's *E-examples from Principles and Standards for School Mathematics* [2012] for more examples of complex problems that require students to apply mathematical practice standards.)

In addition, teachers can spend classroom time discussing complex problems, allowing students to consider the problem-solving approaches other students take and identifying correspondences and differences among the different approaches. This instructional tactic asks students to employ the skills necessary to make sense of problems and persevere in solving them (Mathematical Practice Standard 1).

As students solve multi-step ratio, percent, and proportion problems in context (whether real-world or mathematical), they need to be able to decontextualize the problem—that is, first restate the problem symbolically with numbers, ratios, proportions, or equations and then know how to manipulate those symbols in this new mathematical context. Students also strengthen their ability to solve ratio, percent, and proportion problems if they are also able to contextualize—that is, to see numbers as having a meaning in the context of the problem. The ability to reason both abstractly (to contextualize) and quantitatively (to decontextualize) is described in Mathematical Practice Standard 2.

Applying the concepts of ratios and proportions to problems arising in everyday life, society, and the workplace is related to Mathematical Practice Standard 4, "Model with mathematics." In the middle grades, for example, students might "apply proportional reasoning to plan a school event or analyze a problem in the community" (CCSSI, 2010c, p. 7). The teacher might pose real-world problems or ask the students to formulate problems that they find interesting.

Within the Ratios and Proportional Relationships domain, students are also asked to describe relationships using ratio and rate language. The ability to communicate with precision, using clear definitions, is related to Mathematical Practice Standard 6 ("Attend to precision"). When comparing ratios using tables, students will look for and make use of structure, meeting the requirements of Mathematical Practice Standard 7.

Conceptual Pathway Through the Grades

The Ratios and Proportional Relationships domain is first introduced in 6th grade, but the foundational concepts students need to master this content are presented earlier. Figure 2.2 traces their development through the elementary grades. Teachers preparing instruction on ratios and proportions may find this chart a useful tool to help diagnose and address the readiness levels of their students.

Figure 2.2	Ratios and Proportional Relationships: Conceptual Pathway to Middle School
Grade Level	**Concepts**
Grade 2	• Grouping of equal objects • Partitioning shapes • Fractional terms
Grades 3 and 4	• Multiplication and division of whole numbers • Comparison, addition, and subtraction of fractions with the same denominator • Multiplication of fractions by a whole number
Grade 5	• Multiplication of fractions and division of unit fractions • Addition and subtraction of fractions with different denominators

Now that we've provided an overview of the domain, we'll explore the standards within each grade level.

Grade 6

As noted, the Common Core standards document identifies the development of ratio and proportional reasoning as a critical area for 6th grade. In addition, the development of ratios and proportions is highlighted by both the PARCC frameworks (PARCC, 2011) and the SBAC content specifications for summative assessment of the Common Core mathematics standards

(Schoenfeld, Burkhardt, Abedi, Hess, & Thurlow, 2012). As noted, critical areas provide educators with a place to focus their efforts, minimizing the risk that they will take on too much and, in doing so, limit their ability to address the substantive changes in curriculum the Common Core standards may require.

Understand ratio concepts and use ratio reasoning to solve problems

The development of ratio and proportional reasoning begins with the introduction of ratio relationships, the single cluster in this domain (see Figure 2.3).

Figure 2.3 │ Understand Ratio Concepts and Use Ratio Reasoning to Solve Problems   **6.RP.A**

1. Understand the concept of a ratio and use ratio language to describe a ratio relationship between two quantities. *For example, "The ratio of wings to beaks in the bird house at the zoo was 2:1, because for every 2 wings there was 1 beak." "For every vote candidate A received, candidate C received nearly three votes."*

2. Understand the concept of a unit rate a/b associated with a ratio $a{:}b$ with $b \neq 0$, and use rate language in the context of a ratio relationship. *For example, "This recipe has a ratio of 3 cups of flour to 4 cups of sugar, so there is 3/4 cup of flour for each cup of sugar." "We paid $75 for 15 hamburgers, which is a rate of $5 per hamburger."* (Note: Expectations for unit rates in this grade are limited to non-complex fractions.)

3. Use ratio and rate reasoning to solve real-world and mathematical problems, e.g., by reasoning about tables of equivalent ratios, tape diagrams, double number line diagrams, or equations.

   a. Make tables of equivalent ratios relating quantities with whole-number measurements, find missing values in the tables, and plot the pairs of values on the coordinate plane. Use tables to compare ratios.

   b. Solve unit rate problems including those involving unit pricing and constant speed. *For example, if it took 7 hours to mow 4 lawns, then at that rate, how many lawns could be mowed in 35 hours? At what rate were lawns being mowed?*

   c. Find a percent of a quantity as a rate per 100 (e.g., 30% of a quantity means 30/100 times the quantity); solve problems involving finding the whole, given a part and the percent.

   d. Use ratio reasoning to convert measurement units; manipulate and transform units appropriately when multiplying or dividing quantities.

The introduction to the 6th grade mathematics standards (CCSSI, 2010c, p. 39) provides further insight into the intent behind the statement in Standard 1 (6.RP.A.1) that students will "Understand the concept of a ratio." It briefly describes the desired connections among ratios and rates and a student's ability to reason with multiplication and division, presenting equivalent ratios and rates as extending pairs of rows or columns in a multiplication table. If you are not familiar with the use of the multiplication table as a way to derive equivalent ratios, here's an illustration:

1	2	3	4
2	4	6	8
3	6	9	12
4	8	12	16

Equivalent ratios can be generated simply by juxtaposing rows in the table. For example, pairing the first and second rows results in the ratios 1/2, 2/4, 3/6, and 4/8. Pairing other rows, such as the second and fourth rows, produces similar results, and pairing columns is equally effective.

This demonstration is an easy way to link the new content presented in the 6th grade domain of Ratios and Proportional Relationships to students' prior knowledge of whole-number multiplication, developed in the 3rd and 4th grade standards of the Operations and Algebraic Thinking domain (3.OA, 4.OA), and to their ability to perform operations on fractions, developed and practiced in 5th grade (see Figure 2.2).

In addition to advising teachers to forge a connection between multiplication tables and equivalent fractions, the introduction to this domain also indicates that students should spend some time developing their understanding of ratios and rates through the use of simple drawings that indicate the relative size of quantities. "Simple drawings" can be as uncomplicated as geometric shapes (e.g., circles, rectangles) that show equivalent fractions or as complicated as double number line diagrams representing proportions with different units. As students use these pictures, they should be reminded of their previous knowledge of fractions, obtained formally beginning in grade 3, which gives them an entry point to understanding ratios and rates.

These connections from grade to grade are supplemented by connections across the 6th grade mathematical domains. As 6th graders begin to analyze proportional relationships, they may find it helpful to connect ratios with graphing points in the coordinate plane, as described in the Number System domain (6.NS.C.8). As they use ratios and ratio reasoning to solve unit rate problems or percent problems, they may use variables to represent the unknowns in the problems, which naturally connects with the Expressions and Equations domain (6.EE.A.2, 6.EE.B.6).

Grade 7

As noted, the standards document (CCSSI, 2010c), the PARCC frameworks document (PARCC, 2011), and the SBAC content specifications document (Schoenfeld et al., 2012) all identify the development of ratio and proportional reasoning as a critical area for 7th grade. Seventh graders are asked to build on the ratio concepts introduced in the 6th grade, extending their understanding to include the computation of unit rates and proportional relationships.

At the 7th grade level, this domain's standards are also contained in a single cluster.

Analyze proportional relationships and use them to solve real-world and mathematical problems

The three standards in the cluster (see Figure 2.4) focus on helping students understand direct proportional relationships and reason with proportions.

While the terms *direct* and *inverse proportion* do not appear in the standards document, both the examples included in these standards and the text found in the draft *Progressions for the Common Core State Standards in Mathematics* (Common Core State Standards Writing Team, 2011) suggest

Figure 2.4 | **Analyze Proportional Relationships and Use Them to Solve Real-World and Mathematical Problems**

1. Compute unit rates associated with ratios of fractions, including ratios of lengths, areas and other quantities measured in like or different units. *For example, if a person walks 1/2 mile in each 1/4 hour, compute the unit rate as the complex fraction ½/¼ miles per hour, equivalently 2 miles per hour.*
2. Recognize and represent proportional relationships between quantities.
    a. Decide whether two quantities are in a proportional relationship, e.g., by testing for equivalent ratios in a table or graphing on a coordinate plane and observing whether the graph is a straight line through the origin.
    b. Identify the constant of proportionality (unit rate) in tables, graphs, equations, diagrams, and verbal descriptions of proportional relationships.
    c. Represent proportional relationships by equations. *For example, if total cost* t *is proportional to the number* n *of items purchased at a constant price* p, *the relationship between the total cost and the number of items can be expressed as* t = pn.
    d. Explain what a point (x, y) on the graph of a proportional relationship means in terms of the situation, with special attention to the points $(0, 0)$ and $(1, r)$ where r is the unit rate.
3. Use proportional relationships to solve multi-step ratio and percent problems. *Examples: simple interest, tax, markups and markdowns, gratuities and commissions, fees, percent increase and decrease, percent error.*

that students in 7th grade should be focusing on direct proportions. In 8th grade, students are expected to recognize distance–time equations presented in different equivalent forms as a proportional relationship (8.EE.B.5), which suggests that 8th graders will be working with inverse proportions.

In 7th grade, students begin their work with directly proportional relationships by computing unit rates, as stated in Standard 1 (7.RP.A.1) and go on to represent proportional relationships with equations and explain what a given point on a graph means in terms of a given situation, the topic of Standard 2 (7.RP.A.2). Mastery of these two standards ensures a foundational understanding of proportions that students need in order to solve both single-step and multi-step problems involving ratios and proportions, addressed in Standard 3 (7.RP.A.3).

Students' understanding of proportional relationships will also allow them to make connections to content beyond this domain, notably to relate corresponding lengths between objects and derive the relationship between the circumference and area of a circle—content found in the Geometry domain (7.G.A.1, 7.G.B.4). Although the standards document does not explicitly state that one standard or cluster should be taught prior to another, the logical sequence described here indicates that there is such a sequence—one that will be of special interest to those developing curriculum guides.

For an example of a lesson addressing 7.RP.A.2, please see **Sample Lesson 2**.

As students identify the constant of proportionality in graphs, they will develop an informal understanding of unit rate as a measure of the slope of a line. They'll go on to formalize this understanding in 8th grade, as they learn about the connections among proportional relationships, lines, and linear equations (8.EE.B). Ultimately, this understanding of ratios and proportions will enable students to grasp concepts introduced in high school algebra, statistics, and geometry.

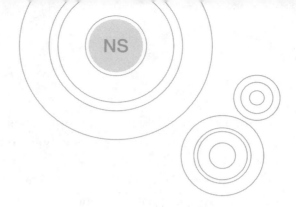

NS

CHAPTER 3

The Number System

The second domain in the Common Core middle school mathematics standards is the Number System (NS), which is found in grades 6 through 8. The standards in this domain build on concepts related to base 10 and fractions, addressed in the elementary-level standards, and they include topics related to operations and rational numbers. Figure 3.1 shows an overview of the clusters within the Number System domain, organized by grade level.

Our examination of the Number System domain begins with a look at the associated Standards for Mathematical Practice and goes on to consider how the content within each cluster builds and connects throughout the middle school grades.

Connections to the Standards for Mathematical Practice

Students learning about the number system will engage in several important mathematical practices. As they begin to make sense of numbers such as fractions or negative integers, in part by using visual fraction models or number lines, they are asked to solve problems involving number sense. Participating in these activities allows students to "make sense

Figure 3.1	The Number System Domain: Middle School Overview	
Grade Level	**Clusters**	**Standards**
Grade 6	**6.NS.A** Apply and extend previous understandings of multiplication and division to divide fractions by fractions.	6.NS.A.1
	6.NS.B Compute fluently with multi-digit numbers and find common factors and multiples.	6.NS.B.2, 6.NS.B.3, 6.NS.B.4
	6.NS.C Apply and extend previous understandings of numbers to the system of rational numbers.	6.NS.C.5, 6.NS.C.6, 6.NS.C.7, 6.NS.C.8
Grade 7	**7.NS.A** Apply and extend previous understandings of operations with fractions to add, subtract, multiply, and divide rational numbers.	7.NS.A.1, 7.NS.A.2, 7.NS.A.3
Grade 8	**8.NS.A** Know that there are numbers that are not rational, and approximate them by rational numbers.	8.NS.A.1, 8.NS.A.2

of problems and persevere in solving them" (Mathematical Practice Standard 1).

Whether students are working with fractions or negative integers to solve problems, they'll benefit from the ability to shift perspectives on a mathematical problem, a skill associated with Mathematical Practice Standard 2, "Reason abstractly and quantitatively." They will pursue this standard further when they interpret sums of rational numbers or solve word problems involving fractions, which involves both thinking about the problem context (e.g., determining how much chocolate each person will get if three people share a half pound of chocolate equally) and thinking about the expression in terms of the numbers that need to be manipulated (e.g., ½ ÷ 3 = ?). Similarly, when students describe contextual situations in which opposite quantities combine to make zero, they're viewing problems in context and as mathematical terms. These skills also relate to Mathematical Practice Standard 4, "Model with mathematics."

To write, interpret, or explain mathematical concepts, students need to be able to communicate their ideas precisely (Mathematical Practice Standard 6). Finally, as they analyze specific examples to find patterns and to develop a sense of the overall structure of the rational number system, students will be doing work connected to Mathematical Practice Standard 7, "Look for and make use of structure."

Conceptual Pathway Through the Grades

While study of the Number System domain begins in 6th grade, work to lay the foundation for understanding the rational number system begins in the first years of schooling. Figure 3.2 traces two key ideas—base ten and fractions—through the elementary grades.

Figure 3.2	**The Number System: Conceptual Pathway to Middle School**		
Grade Level	**Concepts (Base Ten)**	**Concepts (Fractions)**
Kindergarten	• The decomposition of numbers into 10 ones and some further ones	—
Grade 1	• Place value (whole numbers to the tens place) • The use of place value understanding to add and subtract within 100	—
Grade 2	• Place value (whole numbers to the hundreds place) • The use of place value understanding to add and subtract within 1,000	—
Grade 3	• The use of the place value system to perform multi-digit arithmetic (add/subtract within 1,000; multiply one-digit whole numbers by multiples of 10)	• Fractions as numbers (equivalence, place on a number line)

Figure 3.2	**The Number System: Conceptual Pathway to Middle School** *(continued)*	
Grade Level	**Concepts (Base Ten)**	**Concepts (Fractions)**
Grade 4	• The use of the place value system to read, write, round, or compare numbers • The use of the place value system to add and subtract multi-digit numbers; multiply numbers of up to four digits by a one-digit number or multiply two two-digit numbers; or divide whole numbers of up to four digits • Decimal notation for fractions	• Ordering fractions • Adding unit fractions • Multiplying fractions by a whole number
Grade 5	• The use of the place value system to read, write, compare, or round decimals to the thousandths • The use of the place value system to perform operations on decimals to the hundredths	• Adding and subtracting fractions with unlike denominators • Multiplying fractions by whole numbers or fractions • Dividing fractions by whole numbers and whole numbers by fractions • Interpreting multiplication as scaling

We will be referring to this summary of the conceptual pathway for the Number System domain as we describe the connections within the standards for each cluster. Teachers may also find this summary a useful tool for determining the readiness level of students for instruction on the number system.

Now that we've reviewed the domain as a whole, let's look at the clusters and standards at the different grade levels.

Grade 6

At the 6th grade level, the Number System domain contains eight standards, grouped into three clusters.

Apply and extend previous understandings of multiplication and division to divide fractions by fractions

This first cluster, consisting of a single standard (see Figure 3.3), contains content identified by the writers of the Common Core as a critical area that should be a major focus of instruction.

6.NS.A

Figure 3.3 | Apply and Extend Previous Understandings of Multiplication and Division to Divide Fractions by Fractions

1. Interpret and compute quotients of fractions, and solve word problems involving division of fractions by fractions, e.g., by using visual fraction models and equations to represent the problem. *For example, create a story context for (2/3) ÷ (3/4) and use a visual fraction model to show the quotient; use the relationship between multiplication and division to explain that (2/3) ÷ (3/4) = 8/9 because 3/4 of 8/9 is 2/3. (In general, (a/b) ÷ (c/d) = ad/bc.) How much chocolate will each person get if 3 people share 1/2 lb of chocolate equally? How many 3/4-cup servings are in 2/3 of a cup of yogurt? How wide is a rectangular strip of land with length 3/4 mi and area 1/2 square mi?*

As is typical in the Common Core standards, Standard 1 (6.NS.A.1) of the Number System domain for grade 6 is phrased to place less emphasis on the new skill it describes—the division of a fraction by a fraction, which can be taught with a simple algorithm—than it does on ensuring students make sense of the skill's underlying concept. They are asked to draw on both their previous experiences with fractions and their grasp on the operations of multiplication and division to divide fractions by fractions. This understanding of division with fractions is a natural extension of work done in 5th grade, which includes dividing fractions by whole numbers and whole numbers by fractions (5.NF.B.7), and it will help students when they begin to perform operations with rational numbers in 7th grade. The ability to solve word problems involving operations with fractions is also developed in both 4th and 5th grades (4.NF.B.4, 5.NF.B.3, 5.NF.B.6).

Compute fluently with multi-digit numbers and find common factors and multiples

Cluster B focuses on computational fluency and common factors and multiples (see Figure 3.4).

Figure 3.4 | Compute Fluently with Multi-digit Numbers and Find Common Factors and Multiples

6.NS.B

2. Fluently divide multi-digit numbers using the standard algorithm.
3. Fluently add, subtract, multiply, and divide multi-digit decimals using the standard algorithm for each operation.
4. Find the greatest common factor of two whole numbers less than or equal to 100 and the least common multiple of two whole numbers less than or equal to 12. Use the distributive property to express a sum of two whole numbers 1–100 with a common factor as a multiple of a sum of two whole numbers with no common factor. *For example, express 36 + 8 as 4(9 + 2).*

The concepts found in Standards 2, 3, and 4 (6.NS.B.2–4) are not identified as critical areas within the Common Core standards. As such, they are not recommended as the focus of extensive instruction in the 6th grade classroom, but teachers should still spend some time on these concepts, as mastery will help students work more effectively with equations, ratios, and geometric formulas.

The groundwork for understanding how to perform operations on multi-digit numbers is laid in the elementary grades. Beginning in 1st grade, students perform computations with multi-digit numbers, using place value understanding, the properties of operations, and common algorithms. In 3rd grade, they are introduced to the addition and subtraction algorithms for whole numbers less than 1,000 and, in 4th grade, to algorithms for any multi-digit number. The multiplication algorithm is introduced in 5th grade, and the division algorithm is introduced for the first time here, in 6th grade.

The final standard within this cluster, Standard 4 (6.NS.B.4), focuses on finding greatest common factors and least common multiples for sets of numbers—two skills that are commonly found in traditional state standards documents. The Common Core takes a slightly different approach, however, by tying these familiar skills directly to the algebraic concept of factoring. Finding the common factor of the sum of two whole numbers will prepare students for factoring algebraic equations, which is part of the focus of 7.EE.A.1 within the Expressions and Equations domain (see Chapter 4, p. 45). Factoring is first introduced as a topic in 4th grade, when students are asked to find factor pairs and to determine if numbers are prime or composite (4.OA.B.4).

Apply and extend previous understandings of numbers to the system of rational numbers

The third and final cluster of the Number System domain at the 6th grade level introduces the concept of rational numbers (see Figure 3.5). Because students' knowledge of rational numbers—particularly of negative rational numbers—is identified as a critical area in the Common Core standards, this cluster is another that should receive particular emphasis in the classroom.

6.NS.C

Figure 3.5 | **Apply and Extend Previous Understandings of Numbers to the System of Rational Numbers**

5. Understand that positive and negative numbers are used together to describe quantities having opposite directions or values (e.g., temperature above/below zero, elevation above/below sea level, credits/debits, positive/negative electric charge); use positive and negative numbers to represent quantities in real-world contexts, explaining the meaning of 0 in each situation.

6. Understand a rational number as a point on the number line. Extend number line diagrams and coordinate axes familiar from previous grades to represent points on the line and in the plane with negative number coordinates.

    a. Recognize opposite signs of numbers as indicating locations on opposite sides of 0 on the number line; recognize that the opposite of the opposite of a number is the number itself, e.g., $-(-3) = 3$, and that 0 is its own opposite.

    b. Understand signs of numbers in ordered pairs as indicating locations in quadrants of the coordinate plane; recognize that when two ordered pairs differ only by signs, the locations of the points are related by reflections across one or both axes.

    c. Find and position integers and other rational numbers on a horizontal or vertical number line diagram; find and position pairs of integers and other rational numbers on a coordinate plane.

Figure 3.5 | **Apply and Extend Previous Understandings of Numbers to the System of Rational Numbers** *(continued)*

7. Understand ordering and absolute value of rational numbers.
   a. Interpret statements of inequality as statements about the relative position of two numbers on a number line diagram. *For example, interpret –3 > –7 as a statement that –3 is located to the right of –7 on a number line oriented from left to right.*
   b. Write, interpret, and explain statements of order for rational numbers in real-world contexts. *For example, write –3°C > –7°C to express the fact that –3°C is warmer than –7°C.*
   c. Understand the absolute value of a rational number as its distance from 0 on the number line; interpret absolute value as magnitude for a positive or negative quantity in a real-world situation. *For example, for an account balance of –30 dollars, write |–30| = 30 to describe the size of the debt in dollars.*
   d. Distinguish comparisons of absolute value from statements about order. *For example, recognize that an account balance less than –30 dollars represents a debt greater than 30 dollars.*
8. Solve real-world and mathematical problems by graphing points in all four quadrants of the coordinate plane. Include use of coordinates and absolute value to find distances between points with the same first coordinate or the same second coordinate.

Students begin exploring numbers in kindergarten. They learn about counting numbers and then fractions in 3rd grade (3.NF.A) and encounter decimals for the first time in 4th grade (4.NF.C). In 6th grade, students expand their conception of what a number is to include negative numbers, which prepares them for the work of Standard 8 (6.NS.C.8): beginning to graph in all four quadrants of the coordinate plane and incorporating an understanding of the coordinate plane, developed in 5th grade (5.G.A.). The standards in this cluster ask students to understand what a negative number represents, the ordering and absolute value of rational numbers, and the use of negative numbers in graphing. Graphing of rational numbers may be useful as students use the coordinates of the vertices of polygons to find the length of a side in the context of a real-world problem (see Chapter 6's discussion of 6.G.A.3, p. 62).

Grade 7

When students reach 7th grade, the focus of the Number System domain turns to putting their new understanding of rational numbers to work. Extending and using the arithmetic of rational numbers is identified by the Common Core standards document as a critical area, and it is content highlighted by both the PARCC and the SBAC content documents (PARCC, 2011; Schoenfeld et al., 2012).

At this grade level, the domain contains three standards in one cluster.

Apply and extend previous understandings of operations with fractions to add, subtract, multiply, and divide rational numbers

The Number System standards for 7th grade ask students to understand and perform operations on all rational numbers (see Figure 3.6).

7.NS.A

Figure 3.6 | **Apply and Extend Previous Understandings of Operations with Fractions to Add, Subtract, Multiply, and Divide Rational Numbers**

1. Apply and extend previous understandings of addition and subtraction to add and subtract rational numbers; represent addition and subtraction on a horizontal or vertical number line diagram.
   a. Describe situations in which opposite quantities combine to make 0.
      For example, a hydrogen atom has 0 charge because its two constituents are oppositely charged.
   b. Understand $p + q$ as the number located a distance $|q|$ from p, in the positive or negative direction depending on whether q is positive or negative. Show that a number and its opposite have a sum of 0 (are additive inverses). Interpret sums of rational numbers by describing real-world contexts.
   c. Understand subtraction of rational numbers as adding the additive inverse, $p - q = p + (-q)$. Show that the distance between two rational numbers on the number line is the absolute value of their difference, and apply this principle in real-world contexts.
   d. Apply properties of operations as strategies to add and subtract rational numbers.

Figure 3.6 | **Apply and Extend Previous Understandings of Operations with Fractions to Add, Subtract, Multiply, and Divide Rational Numbers** *(continued)*

2. Apply and extend previous understandings of multiplication and division and of fractions to multiply and divide rational numbers.
   a. Understand that multiplication is extended from fractions to rational numbers by requiring that operations continue to satisfy the properties of operations, particularly the distributive property, leading to products such as $(-1)(-1) = 1$ and the rules for multiplying signed numbers. Interpret products of rational numbers by describing real-world contexts.
   b. Understand that integers can be divided, provided that the divisor is not zero, and every quotient of integers (with non-zero divisor) is a rational number. If p and q are integers, then $-(p/q) = (-p)/q = p/(-q)$. Interpret quotients of rational numbers by describing real-world contexts.
   c. Apply properties of operations as strategies to multiply and divide rational numbers.
   d. Convert a rational number to a decimal using long division; know that the decimal form of a rational number terminates in 0s or eventually repeats.
3. Solve real-world and mathematical problems involving the four operations with rational numbers. (*Note*: Computations with rational numbers extend the rules for manipulating fractions to complex fractions.)

While operations with rational numbers (particularly negative rational numbers) can be taught as "rules" (e.g., "When multiplying, two negatives equal a positive"), the Common Core standards require that students acquire a deep understanding of how the properties of operations extend to rational numbers. The first standard here, Standard 1 (7.NS.A.1), asks students to "apply and extend previous understandings of addition and subtraction to add and subtract rational numbers." The previous understandings referenced include grasping the meaning of and the relationship between addition and subtraction and having a working knowledge of the properties of operations, both of which stem from students' first encounters with the concepts of addition and subtraction in kindergarten through 3rd grade (see Figure 3.2).

In these earliest grades, students learn what addition and subtraction represent and use the relationship between the two operations and what they know about place value to add and subtract whole numbers (K.OA.A, 1.OA.A–D, 2.OA.A–C). In 3rd and 4th grades, students become fluent with the standard algorithm for addition and subtraction (3.NBT.A.2, 4.NBT.B.4). Then, in 4th through 6th grades, they apply their understanding of the properties of addition and subtraction to add and subtract fractions and decimals (4.NF.B.3, 4.NF.C.5, 5.NF.A.1, 5.NBT.B.7, 6.NS.B.3). Sixth grade is also when students are introduced to the concept of positive and negative numbers, with an emphasis on understanding negative and positive numbers as points on a number line or graph (6.NS.C). All these understandings will help 7th graders grasp opposite quantities, understand subtraction as adding the additive inverse, and apply properties of operations as strategies to add and subtract rational numbers.

Standard 2 (7.NS.A.2) is similar to its predecessor but focuses on different operations. It asks students to draw on previous understandings of multiplication, division, and fractions in order to understand how to multiply and divide rational numbers. These previous understandings are first established in 3rd grade, when students learn to multiply and divide whole numbers (3.OA.A–D). Third grade also introduces the properties of multiplication and division, along with place value understanding (3.NBT.A). In 4th grade, students begin to multiply multi-digit numbers using strategies based on place value and the properties of operations (4.NBT.B.5), and by 5th grade, they are using the properties of operations to multiply fractions (5.NF.B.4–6). In 6th grade, students use their understandings of the properties of operations to divide fractions by fractions (6.NS.A.1). The Common Core standards' intention is that as students develop a deeper understanding of *why* the rules for the multiplication and division of negative rational numbers reflect the properties of operations, they will be more likely to apply those rules properly.

Grade 8

At the 8th grade level, the two standards in the Number System domain's sole cluster introduce students to the concept of irrational numbers. This concept is not identified by the Common Core standards as a critical area. However, as it helps students master the critical skill of solving problems involving the Pythagorean Theorem, it may be thought of as an important supporting concept. Although students must understand irrational numbers in order to do high school mathematics, the 8th grade Number

System standards are ones to address succinctly and are not recommended as the focus of extensive instruction.

Know that there are numbers that are not rational, and approximate them by rational numbers

As noted, the two 8th grade standards in the Number System domain both deal with irrational numbers (see Figure 3.7).

Figure 3.7 | **Know That There Are Numbers That Are Not Rational, and Approximate Them by Rational Numbers**

8.NS.A

1. Know that numbers that are not rational are called irrational. Understand informally that every number has a decimal expansion; for rational numbers show that the decimal expansion repeats eventually, and convert a decimal expansion which repeats eventually into a rational number.
2. Use rational approximations of irrational numbers to compare the size of irrational numbers, locate them approximately on a number line diagram, and estimate the value of expressions (e.g., π^2). *For example, by truncating the decimal expansion of $\sqrt{2}$, show that $\sqrt{2}$ is between 1 and 2, then between 1.4 and 1.5, and explain how to continue on to get better approximations.*

Standard 1 (8.NS.A.1) asks students to look beyond the boundaries of rational numbers and explore the concept of irrational numbers. In 7th grade, students learn the formulas for the area and circumference of a circle and how to derive the relationship between the area and circumference of a circle; in the process, they are necessarily introduced to π (7.G.B.4). Teachers can use students' background knowledge of π as an entry point for understanding rational numbers.

An understanding of how rational numbers might be connected to familiar concepts will be helpful to students as they tackle Standard 2 (8.NS.A.2), which asks them to use rational approximations of irrational numbers. Eighth grade students will also need to incorporate their understanding of irrational numbers when finding solutions to equations of the form $x^2 = p$ (8.EE.A.2) and when applying the Pythagorean Theorem to solve problems (8.G.B.7).

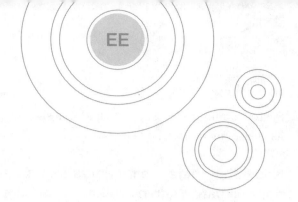

CHAPTER 4

Expressions and Equations

The introduction to the Common Core standards' high school Algebra domain provides some useful definitions to help differentiate the two mathematical terms *expression* and *equation*:

> An expression is a record of a computation with numbers, symbols that represent numbers, arithmetic operations, exponentiation, and, at more advanced levels, the operation of evaluating a function.... An equation is a statement of equality between two expressions, often viewed as a question asking for which values of the variables the expressions on either side are in fact equal. These values are the solutions to the equation. (CCSSI, 2010c, p. 62)

The middle school Expressions and Equations (EE) domain provides a critical bridge between content in the Operations and Algebraic Thinking domain in earlier grades and algebraic content students will encounter in high school. Figure 4.1 shows an overview of the Expressions and Equations domain's clusters and standards by grade level.

Taken together, the standards in this domain are designed to help students understand, create, and manipulate expressions, equations, and inequalities.

Figure 4.1	**The Expressions and Equations Domain: Middle School Overview**	
Grade Level	**Clusters**	**Standards**
Grade 6	**6.EE.A** Apply and extend previous understandings of arithmetic to algebraic expressions.	6.EE.A.1, 6.EE.A.2, 6.EE.A.3, 6.EE.A.4
	6.EE.B Reason about and solve one-variable equations and inequalities.	6.EE.B.5, 6.EE.B.6, 6.EE.B.7, 6.EE.B.8
	6.EE.C Represent and analyze quantitative relationships between dependent and independent variables.	6.EE.C.9
Grade 7	**7.EE.A** Use properties of operations to generate equivalent expressions.	7.EE.A.1, 7.EE.A.2
	7.EE.B Solve real-life and mathematical problems using numerical and algebraic expressions and equations.	7.EE.B.3, 7.EE.B.4
Grade 8	**8.EE.A** Work with radicals and integer exponents.	8.EE.A.1, 8.EE.A.2, 8.EE.A.3, 8.EE.A.4,
	8.EE.B Understand the connections between proportional relationships, lines, and linear equations.	8.EE.B.5, 8.EE.B.6
	8.EE.C Analyze and solve linear equations and pairs of simultaneous linear equations.	8.EE.C.7, 8.EE.C.8

Connections to the Standards for Mathematical Practice

Students will use several of the Standards for Mathematical Practice as they learn to work with expressions and equations.

Relating the properties of operations of numbers to algebraic expressions helps students gain a familiarity with expressions and equations that will allow them to use expressions and equations to solve problems within real-world contexts. Teachers can challenge students by creating problems that engage them in the analysis of givens, constraints, relationships, and goals, activities that reflect an important aspect of Mathematical Practice Standard 1, "Make sense of problems and persevere in solving them."

As students solve real-world contextual problems, their ability to understand and represent a problem both in algebraic notation and in context will provide evidence that they can "reason abstractly and quantitatively" (Mathematical Practice Standard 2) and "model with mathematics" (Mathematical Practice Standard 4).

When students use stated assumptions, definitions, and previously established mathematical tools to demonstrate the derivation of an equation, such as the equation of a line, they are honing skills related to Mathematical Practice Standard 3, "Construct viable arguments," and Mathematical Practice Standard 6, "Attend to precision."

Their strategic use of tools (Mathematical Practice Standard 5), such as the use of physical models or computer software to solve multi-step problems that include positive and negative numbers, will help students understand the problems and the steps required to find solutions.

Conceptual Pathway Through the Grades

While the Expressions and Equations domain begins in 6th grade, the foundational ideas necessary for understanding expressions and equations are introduced throughout the elementary grades, as illustrated in Figure 4.2.

Figure 4.2	**Expressions and Equations: Conceptual Pathway to Middle School**	
Grade Level	**Concepts**
Kindergarten	• Understanding addition as putting together and adding to • Understanding subtraction as taking apart and taking from
Grade 1	• Representation of addition and subtraction problems using numbers less than 20 with symbols for the unknown number • Properties of operations • The relationship between addition and subtraction
Grade 2	• Representation of addition and subtraction problems with symbols for the unknown number

Figure 4.2	**Expressions and Equations: Conceptual Pathway to Middle School** *(continued)*	
Grade Level	**Concepts**	
Grade 3	• Representation of multiplication and division problems using symbols for the unknown number • Properties of multiplication and the relationship between multiplication and division • Problems involving the four operations • Patterns in arithmetic	
Grade 4	• Multiplicative and additive comparisons • Word problems using all four operations • Factors and multiples • Patterns	
Grade 5	• Whole-number exponents to denote powers of 10 • Numerical expressions (order of operations, translate equations expressed as words to symbolic form) • Patterns	

This overview provides a summary of the conceptual pathway for this topic. Teachers planning instruction on equations and expressions may also find it a useful tool for helping to determine students' readiness levels.

Let's turn now to the grade-by-grade look at the clusters in this domain and the standards they contain.

Grade 6

The Common Core standards document designates the content described in the Expressions and Equations domain as a critical area in 6th grade. The content here is also highlighted by both the *PARCC Model Content Frameworks* (PARCC, 2011) and SBAC's *Content Specifications for the Summative Assessment of the Common Core State Standards for Mathematics* (Schoenfeld et al., 2012). Remember, critical areas point out where educators should focus their efforts in the hope of preventing them from taking on too much and struggling to implement the substantive changes that the Common Core standards require.

At the 6th grade level, the Expressions and Equations domain contains nine standards, organized into three clusters.

Apply and extend previous understandings of arithmetic to algebraic expressions

The development of expressions and equations begins with the connection of arithmetic to algebraic expressions. The standards in Cluster A (see Figure 4.3) focus on using exponents; reading, writing, and evaluating expressions; and generating equivalent expressions.

Standard 1 (6.EE.A.1) asks students to write and evaluate expressions involving exponents. Students are first introduced to exponents in in 5th grade, when they explore the place value system and learn how to denote powers of 10 (5.NBT.A.2). This understanding is extended in 6th

6.EE.A

Figure 4.3 | **Apply and Extend Previous Understandings of Arithmetic to Algebraic Expressions**

1. Write and evaluate numerical expressions involving whole-number exponents.
2. Write, read, and evaluate expressions in which letters stand for numbers.
   a. Write expressions that record operations with numbers and with letters standing for numbers. *For example, express the calculation "Subtract y from 5" as 5 – y.*
   b. Identify parts of an expression using mathematical terms (sum, term, product, factor, quotient, coefficient); view one or more parts of an expression as a single entity. *For example, describe the expression 2(8 + 7) as a product of two factors; view (8 + 7) as both a single entity and a sum of two terms.*
   c. Evaluate expressions at specific values of their variables. Include expressions that arise from formulas used in real-world problems. Perform arithmetic operations, including those involving whole-number exponents, in the conventional order when there are no parentheses to specify a particular order (Order of Operations). *For example, use the formulas V = s³ and A = 6s² to find the volume and surface area of a cube with sides of length s = 1/2.*
3. Apply the properties of operations to generate equivalent expressions. *For example, apply the distributive property to the expression 3(2 + x) to produce the equivalent expression 6 + 3x; apply the distributive property to the expression 24x + 18y to produce the equivalent expression 6(4x + 3y); apply properties of operations to y + y + y to produce the equivalent expression 3y.*
4. Identify when two expressions are equivalent (i.e., when the two expressions name the same number regardless of which value is substituted into them). *For example, the expressions y + y + y and 3y are equivalent because they name the same number regardless of which number y stands for.*

grade, as they begin to write and evaluate numerical expressions involving exponents and use the order of operations on expressions including exponents (6.EE.A.1–2). Students will use these skills again in 8th grade, when they move on to work with integer exponents and equations involving exponents (see 8.EE.A, p. 48), and in high school, when they learn about rational exponents (N-RN.A.1–2).

The content addressed in Standard 2 (6.EE.A.2), learning to read, write, and evaluate expressions, has its beginnings in the early elementary grades, when students represent problems using equations with a symbol standing in for the unknown number (1.OA.A.2, 2.OA.A.1, 3.OA.A.3). In 4th grade, students begin using a letter as a symbol for an unknown (4.OA.A.3), and in 5th grade, they start to interpret numerical expressions, translating between a verbal description of a calculation and its symbolic representation (5.OA.A.2). As students develop a formal understanding of the use of variables in mathematical expressions (as well as equations and inequalities—concepts found in 6.EE.B; see p. 42), the abilities they developed in earlier grades will help them successfully evaluate expressions and generate equivalent expressions.

In the lower grades, as students learn about each operation, they also spend ample time focusing on its properties. Students in kindergarten learn the basic meaning of addition and subtraction while gaining a conceptual understanding of what is meant by each operation (K.OA.A.1–4). In 1st and 2nd grades, they apply the properties of operations as strategies to add and subtract whole numbers. In 3rd and 4th grades, they begin to apply properties of operations as strategies to multiply and divide whole numbers and learn to use the properties of operations to explain arithmetic patterns (3.OA.B, 4.OA.B.5, 4.NBT.B). When students reach 4th grade, they begin to add and subtract fractions using the properties of operations (4.NF.B), and in 5th grade, they use the properties of operations to add, subtract, multiply, and divide decimal numbers (5.NBT.B.6–7). Sixth graders draw upon the numbers skills and concepts introduced in earlier grades to apply the properties of operations as strategies to generate equivalent expressions.

As students apply the properties of operations to generate equivalent expressions, as Standard 3 requires, the ability to find common factors and multiples—a skill found in the Number System domain (6.NS.B.4)—will also come in handy.

Reason about and solve one-variable equations and inequalities

Sixth graders' ability to find common factors and multiples will also be useful as they address the standards in the domain's second cluster, which relates to equations and inequalities that include one variable (see Figure 4.4).

As is often the case in the Common Core standards, the first standard in Cluster B, Standard 5 (6.EE.B.5), stresses conceptual understanding—here, understanding solving equations or inequalities—and establishes the skill involved (substitution) as a secondary goal. The perspective of equations being "true" or "false" should not be a new one; consider that 1st graders are asked both to understand the meaning of the equal sign and to decide if equations are true or false (1.OA.D.7). As students progress through

6.EE.B

Figure 4.4 | **Reason About and Solve One-Variable Equations and Inequalities**

5. Understand solving an equation or inequality as a process of answering a question: which values from a specified set, if any, make the equation or inequality true? Use substitution to determine whether a given number in a specified set makes an equation or inequality true.
6. Use variables to represent numbers and write expressions when solving a real-world or mathematical problem; understand that a variable can represent an unknown number, or, depending on the purpose at hand, any number in a specified set.
7. Solve real-world and mathematical problems by writing and solving equations of the form $x + p = q$ and $px = q$ for cases in which p, q, and x are all nonnegative rational numbers.
8. Write an inequality of the form $x > c$ or $x < c$ to represent a constraint or condition in a real-world or mathematical problem. Recognize that inequalities of the form $x > c$ or $x < c$ have infinitely many solutions; represent solutions of such inequalities on number line diagrams.

the elementary grades, they are required to determine which unknown number will make a given equation true for a variety of operations and numbers. In 4th grade, they are asked to use letters for the unknown quantity (4.OA.A.3). Here in 6th grade, students' understanding of equations or inequalities connects directly to their ability to "write, read, and evaluate expressions in which letters stand for numbers," the focus of the domain's Standard 2 (6.EE.A.2). Once they've learned how variables can be used in an expression, students will find it easier to extend their understanding to equations.

The final standard in Cluster B, Standard 8 (6.EE.B.8), focuses on inequalities. An understanding of inequalities stems from the ability to compare numbers, which is a skill initially introduced in 1st grade (1.NBT.B.3). As students encounter different types of numbers during the elementary grades (larger numbers, fractions, decimals), they are asked to compare these numbers and to represent them on number lines. Now, in 6th grade, students' ability to understand the >, <, and = symbols and variables will allow them to use inequalities to represent constraints within a problem.

Represent and analyze quantitative relationships between dependent and independent variables

Cluster C of the 6th grade Expressions and Equations domain (see Figure 4.5) introduces the concept of dependent and independent variables.

6.EE.C

Figure 4.5 | **Represent and Analyze Quantitative Relationships Between Dependent and Independent Variables**

9. Use variables to represent two quantities in a real-world problem that change in relationship to one another; write an equation to express one quantity, thought of as the dependent variable, in terms of the other quantity, thought of as the independent variable. Analyze the relationship between the dependent and independent variables using graphs and tables, and relate these to the equation. *For example, in a problem involving motion at constant speed, list and graph ordered pairs of distances and times, and write the equation* $d = 65t$ *to represent the relationship between distance and time.*

As you see, this cluster contains just a single standard, but it's a complex one. Standard 9 (6.EE.C.9) covers several concepts: representing two quantities in a real-world problem, writing an equation to represent a relationship between variables, analyzing relationships between variables using graphs and tables, and relating graphs and tables to equations. These ideas can be connected very naturally to the critical area of ratios, as students can use their understanding of ratio relationships to represent and analyze relationships between variables. Connecting these concepts explicitly in lessons may help enhance student understanding of both. To ensure that students do understand, teachers should prompt them to describe their analyses in words as well as with symbolic expressions. The graphs and tables students create may be done by hand or with the help of technology to allow for more flexibility.

Students' capacity to use variables has immediate roots in their capacity to "write, read, and evaluate expressions in which letters stand for numbers" (6.EE.A.2) and can be traced back through the earlier grades (see Figure 4.1). Work to develop the capacity to graph on the coordinate plane begins in 5th grade (5.OA.B.3, 5.G.A), and this skill is addressed in other 6th grade standards as well (6.RP.A.3, 6.NS.C.6, 6.NS.C.8). In 8th grade, students will use the understandings developed in this cluster to work with functions (see Chapter 5).

Grade 7

In 7th grade, students extend their understanding of equivalent expressions and solve real-life and mathematical problems using expressions and equations. The concepts found in this domain are highlighted by both the PARCC frameworks document (PARCC, 2011) and the SBAC content specifications (Schoenfeld et al., 2012).

At the 7th grade level, the Expressions and Equations domain contains four standards, grouped in two clusters.

Use properties of operations to generate equivalent expressions

The standards in Cluster A (see Figure 4.6) are clearly and explicitly connected to an understanding of the properties of operations, which students have been building since kindergarten.

Figure 4.6	**Use Properties of Operations to Generate Equivalent Expressions**

1. Apply properties of operations as strategies to add, subtract, factor, and expand linear expressions with rational coefficients.
2. Understand that rewriting an expression in different forms in a problem context can shed light on the problem and how the quantities in it are related. *For example,* a + 0.05a = 1.05a *means that "increase by 5%" is the same as "multiply by 1.05."*

As discussed earlier, in the lower grades, students spend time focusing on the properties of each mathematical operation as they learn about it. Now, in 7th grade, they are asked to use their understanding of the properties of operations as a problem-solving tool.

As described in Standard 1 (7.EE.A.1), when students add, subtract, factor, and expand linear expressions, they should not be practicing isolated skills but, rather, using strategies that follow directly from what they know about the properties of operations. One example shared in *Progressions for the Common Core State Standards in Mathematics* (Common Core State Standards Writing Team, 2012) is that when students group like terms (e.g., $3x + 12x$), their understanding of the strategies should stem from mastery of the concept of the distributive property (e.g., $[3 + 12]x = 15x$). This overall approach will support students' future work with more complex algebraic equations and expressions.

As is common in the Common Core, Standard 2 (7.EE.A.2) focuses on building a fundamental understanding rather than executing a specific process. In the case of this standard, students are asked to understand the possible utility in rewriting an expression in different forms—a skill described in Standard 1. The example highlights how necessary it is for students to be able to translate symbolic equations into verbal descriptions, a skill introduced in 5th grade (5.OA.A2) and further developed in 6th grade (6.EE.A.2).

Solve real-life and mathematical problems using numerical and algebraic expressions and equations

Cluster B of the 7th grade Expressions and Equations domain (see Figure 4.7) contains content that connects to several other components of the Common Core.

7.EE.B

Figure 4.7 | **Solve Real-Life and Mathematical Problems Using Numerical and Algebraic Expressions and Equations**

3. Solve multi-step real-life and mathematical problems posed with positive and negative rational numbers in any form (whole numbers, fractions, and decimals), using tools strategically. Apply properties of operations to calculate with numbers in any form; convert between forms as appropriate; and assess the reasonableness of answers using mental computation and estimation strategies. *For example: If a woman making $25 an hour gets a 10% raise, she will make an additional 1/10 of her salary an hour, or $2.50, for a new salary of $27.50. If you want to place a towel bar 9¾ inches long in the center of a door that is 27½ inches wide, you will need to place the bar about 9 inches from each edge; this estimate can be used as a check on the exact computation.*

4. Use variables to represent quantities in a real-world or mathematical problem, and construct simple equations and inequalities to solve problems by reasoning about the quantities.

   a. Solve word problems leading to equations of the form $px + q = r$ and $p(x + q) = r$, where p, q, and r are specific rational numbers. Solve equations of these forms fluently. Compare an algebraic solution to an arithmetic solution, identifying the sequence of the operations used in each approach. *For example, the perimeter of a rectangle is 54 cm. Its length is 6 cm. What is its width?*

   b. Solve word problems leading to inequalities of the form $px + q > r$ or $px + q < r$, where p, q, and r are specific rational numbers. Graph the solution set of the inequality and interpret it in the context of the problem. *For example: As a salesperson, you are paid $50 per week plus $3 per sale. This week you want your pay to be at least $100. Write an inequality for the number of sales you need to make, and describe the solutions.*

Looking at Standard 3 (7.EE.B.3), we can immediately see the integration of positive and negative rational numbers with equations, explicitly tying the content to that found in the Number System domain (see Chapter 3). The use of strategic tools links Standard 3 to Mathematical Practice Standard 5, which explains that tools might include pencil and paper, concrete models, or a calculator, among others. As students use these tools to solve multi-step problems, they should be aware of the tools' limitations and use mental computation and estimation strategies to ensure that their final answers are reasonable. As noted, students' use of the properties of operations begins in 3rd grade and expands as they move from grade level to grade level. Standard 3 asks them to apply these properties to calculate with numbers in any form.

The second standard in this cluster, Standard 4 (7.EE.B.4), asks students to represent quantities in real-world or mathematical problems using variables and to construct simple equations (specifically, $px + q = r$ or $p[x + q] = r$) and inequalities ($px + q > r$ or $px + q < r$). These equations are a logical progression from the equations described in 6.EE.C.7–8, the third cluster of the 6th grade standards for this domain ($x + p = q$; $px = q$ and $x > c$; $x < c$), and they will prepare students to use linear equations in 8th grade (8.EE.B, 8.EE.C; see pp. 49–50).

Grade 8

The Common Core standards document identifies formulating and reasoning about expressions and equations as a critical area for 8th grade. This content is also highlighted by both the *PARCC Model Content Frameworks* (PARCC, 2011) and SBAC's *Content Specifications for the Summative Assessment of the Common Core State Standards for Mathematics* (Schoenfeld et al., 2012). As previously noted, critical areas provide educators with a place to focus their efforts, helping to ensure they provide students with sufficient time to develop a deep understanding of essential concepts.

There are eight standards in the Expressions and Equations domain at the 8th grade level, organized into three clusters.

Work with radicals and integer exponents

The 8th grade standards in the Expressions and Equations domain begin with the introduction of radicals and integer exponents (see Figure 4.8).

8.EE.A

Figure 4.8 | **Work with Radicals and Integer Exponents**

1. Know and apply the properties of integer exponents to generate equivalent numerical expressions. *For example, $3^2 \times 3^{-5} = 3^{-3} = 1/3^3 = 1/27$.*
2. Use square root and cube root symbols to represent solutions to equations of the form $x^2 = p$ and $x^3 = p$, where p is a positive rational number. Evaluate square roots of small perfect squares and cube roots of small perfect cubes. Know that $\sqrt{2}$ is irrational.
3. Use numbers expressed in the form of a single digit times an integer power of 10 to estimate very large or very small quantities, and to express how many times as much one is than the other. *For example, estimate the population of the United States as 3×10^8 and the population of the world as 7×10^9, and determine that the world population is more than 20 times larger.*
4. Perform operations with numbers expressed in scientific notation, including problems where both decimal and scientific notation are used. Use scientific notation and choose units of appropriate size for measurements of very large or very small quantities (e.g., use millimeters per year for seafloor spreading). Interpret scientific notation that has been generated by technology.

In 5th grade, students are introduced to exponents as they begin to explore the place value system and understand how to denote powers of 10 (5.NBT.A.2). They extend this understanding in 6th grade as they begin to write and evaluate numerical expressions involving exponents and use the order of operations on expressions including exponents (6.EE.A.1–2, see p. 40). In 8th grade, Standards 1, 2, 3, and 4 (8.EE.A.1–4) introduce students to integer exponents and equations involving exponents. The understanding they gain here will assist them in their high school work with rational exponents (HSN-RN.A.1–2).

Understand the connections between proportional relationships, lines, and linear equations

The second cluster in this domain (see Figure 4.9) explicitly connects the idea of proportional relationships to lines and linear equations.

Figure 4.9 | Understand the Connections Between Proportional Relationships, Lines, and Linear Equations

5. Graph proportional relationships, interpreting the unit rate as the slope of the graph. Compare two different proportional relationships represented in different ways. *For example, compare a distance–time graph to a distance–time equation to determine which of two moving objects has greater speed.*

6. Use similar triangles to explain why the slope *m* is the same between any two distinct points on a non-vertical line in the coordinate plane; derive the equation $y = mx$ for a line through the origin and the equation $y = mx + b$ for a line intercepting the vertical axis at *b*.

Students are first introduced to ratios in 6th grade (6.RP.A), and in 7th grade, they further their understanding of proportions, equations of the form $px + q = r$, and similarity. Connecting these three concepts fosters a deeper understanding the equation of a line ($y = mx + b$), allowing students to connect unit rate to the slope of a graph and use similar triangles to explain why the slope is the same regardless of where on a given line it is measured. Now, in 8th grade, Standard 5 (8.EE.B.5) asks students to apply this understanding of the equation of a line to solve linear equations and to describe the association between two quantities in bivariate data.

In order to derive the equations $y = mx$ and $y = mx + b$, as required in Standard 6 (8.EE.B.6), students must integrate a number of previously acquired skills and knowledge, applying an understanding of the coordinate system and how to manipulate equations. The former is first developed in 5th grade (5.OA.B.3d, 5.G.A.2) and further addressed in 6th grade (6.RP.A.3, 6.NS.C.6, 6.NS.C.8). Writing equivalent expressions is first introduced in 6th grade (6.EE.A.3; see p. 40) and is further developed in 7th grade (7.EE.A). Mathematics teachers may wish to collaborate with science teachers when teaching about unit rate, as science teachers may be using these skills to explain concepts such as force and motion.

Analyze and solve linear equations and pairs of simultaneous linear equations

Both standards within Cluster C (see Figure 4.10) require students to work with linear equations, a skill rooted not only in the previous cluster but

Figure 4.10 | **Analyze and Solve Linear Equations and Pairs of Simultaneous Linear Equations**

7. Solve linear equations in one variable.
   a. Give examples of linear equations in one variable with one solution, infinitely many solutions, or no solutions. Show which of these possibilities is the case by successively transforming the given equation into simpler forms, until an equivalent equation of the form $x = a$, $a = a$, or $a = b$ results (where a and b are different numbers).
   b. Solve linear equations with rational number coefficients, including equations whose solutions require expanding expressions using the distributive property and collecting like terms.
8. Analyze and solve pairs of simultaneous linear equations.
   a. Understand that solutions to a system of two linear equations in two variables correspond to points of intersection of their graphs, because points of intersection satisfy both equations simultaneously.
   b. Solve systems of two linear equations in two variables algebraically, and estimate solutions by graphing the equations. Solve simple cases by inspection. *For example, $3x + 2y = 5$ and $3x + 2y = 6$ have no solution because $3x + 2y$ cannot simultaneously be 5 and 6.*
   c. Solve real-world and mathematical problems leading to two linear equations in two variables. *For example, given coordinates for two pairs of points, determine whether the line through the first pair of points intersects the line through the second pair.*

also in the ability to generate equivalent equations, developed in 6th and 7th grade (6.EE.A.3, 7.EE.A). A solid understanding of how to graph equations, developed in 5th grade (5.OA.B.3, 5.G.A) and further addressed in 6th grade (6.RP.A.3, 6.NS.C.6, 6.NS.C.8), is also very helpful to 8th graders learning to solve linear equations (Standard 7, 8.EE.C.7) and pairs of linear equations (Standard 8, 8.EE.C.8).

Many of the skills associated with the ability to create equations and use them to solve problems are required by standards outside the Equations and Expressions domain. Students will certainly need these skills to successfully build and reason with functions, the topic addressed in the next chapter.

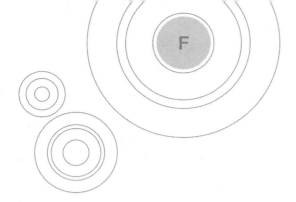

CHAPTER 5

Functions

The Common Core standards take a somewhat unusual approach to the topic of functions. In the traditional approach evidenced in the majority of state standards documents predating the Common Core, functions are not treated as separate subject matter; they are addressed concurrently with expressions and equations in an overarching Algebra strand. By contrast, the Common Core standards present functions as a domain at the 8th grade level of the K–8 mathematics standards and as a stand-alone conceptual category at the high school level. The category's introduction in the high school standards offers the following description of what a function is and how functions are connected to expressions and equations:

> Functions describe situations where one quantity determines another. . . .
> In school mathematics, functions usually have numerical inputs and out-
> puts and are often defined by an algebraic expression. . . . A function
> can be described in various ways, such as by a graph (e.g., the trace
> of a seismograph); by a verbal rule, as in, "I'll give you a state, you give
> me the capital city"; by an algebraic expression like $f(x) = a + bx$; or by
> a recursive rule. . . . Determining an output value for a particular input
> involves evaluating an expression; finding inputs that yield a given output
> involves solving an equation. Questions about when two functions have
> the same value for the same input lead to equations, whose solutions can
> be visualized from the intersection of their graphs. (CCSSI, 2010c, p. 67)

In this chapter, we'll look at the Common Core standards that introduce functions, preparing 8th graders for high school study of functions and modeling. Figure 5.1 shows an overview of the Functions (F) domain at the middle school level.

Figure 5.1	**The Functions Domain: Middle School Overview**		
Grade Level	**Clusters**	**Standards**
Grade 8	**8.F.A** Define, evaluate, and compare functions.	8.F.A.1, 8.F.A.2, 8.F.A.3
	8.F.B Use functions to model relationships between quantities.	8.F.B.4, 8.F.B.5

Connections to the Standards for Mathematical Practice

The Functions domain focuses on conveying a foundational understanding of the topic: teaching students what functions are, what they look like, and the properties of functions regardless of how they're represented (i.e., graphical, algebraic, table, or verbal).

After students begin to make sense of functions, in part by comparing the properties of two functions represented in different ways, they first are challenged to construct a function to model a linear relationship and then are asked to interpret the rate of change and initial value in the context of the situation that it models. The ability to understand and represent a problem as a function is evidence that students are able to "Reason abstractly and quantitatively" (Mathematical Practice Standard 2) and "Model with mathematics" (Mathematical Practice Standard 4).

Students who can describe the functional relationship between two quantities and sketch a graph of a function that has been described verbally are illustrating skills related to Mathematical Practice Standard 6, "Attend to precision." Finally, when students compare the properties of functions represented in different ways, they are required to "Look for and make use of structure" (Mathematical Practice Standard 7), and they begin to see a function both as a whole entity and as a combination of its parts.

Conceptual Pathway Through the Grades

Although the Functions domain is presented for the first time in 8th grade, the concepts necessary for understanding functions are introduced much earlier. Figure 5.2 traces the pathway through the elementary grades.

Figure 5.2	**Functions: Conceptual Pathway to Grade 8**	
Grade Level	**Concepts**
Kindergarten	• Understand addition as putting together and adding to • Understand subtraction as taking apart and taking from
Grade 1	• Representation of addition and subtraction problems up to 20 using symbols for the unknown number • Properties of operations • The relationship between addition and subtraction
Grade 2	• Representation of addition and subtraction problems using symbols for the unknown number
Grade 3	• Representation of multiplication and division problems using symbols for the unknown number • Properties of multiplication and the relationship between multiplication and division • Problems involving the four operations • Patterns in arithmetic
Grades 4	• Multiplicative and additive comparisons • Word problems using all four operations • Factors and multiples • Patterns
Grade 5	• Whole-number exponents to denote powers of 10 • Numerical expressions (order of operations, translate equations expressed as words to symbolic form) • Patterns
Grade 6	• Dependent and independent variables
Grade 7	• Proportional relationships • Simple equations and inequalities

This overview provides a summary of the conceptual pathway for this topic. Teachers may find it a useful tool for assessing students' readiness to learn about functions. We will be referring to this table as we describe the connections within the standards for each cluster.

Grade 8

The development of students' understanding of functions begins with the connection of arithmetic to algebraic expressions. This understanding is identified as a critical area in 8th grade within the Common Core standards, and it is content that is highlighted by the SBAC document (Schoenfeld et al., 2012).

At the middle school level, the Functions domain contains five standards, organized into two clusters.

Define, evaluate, and compare functions

Cluster A focuses on fundamental understandings about functions (see Figure 5.3).

8.F.A

Figure 5.3 | **Define, Evaluate, and Compare Functions**

1. Understand that a function is a rule that assigns to each input exactly one output. The graph of a function is the set of ordered pairs consisting of an input and the corresponding output. (Note: Function notation is not required in Grade 8.)
2. Compare properties of two functions each represented in a different way (algebraically, graphically, numerically in tables, or by verbal descriptions). *For example, given a linear function represented by a table of values and a linear function represented by an algebraic expression, determine which function has the greater rate of change.*
3. Interpret the equation $y = mx + b$ as defining a linear function, whose graph is a straight line; give examples of functions that are not linear. *For example, the function A = s² giving the area of a square as a function of its side length is not linear because its graph contains the points (1, 1), (2, 4) and (3, 9), which are not on a straight line.*

As is often the case in the Common Core standards, Standard 1 (8.F.A.1) focuses on conceptual understanding. As students work to grasp the concept of a function, giving a specific definition and demonstrating

comprehension of the implications, they will benefit from a solid understanding of linear equations and graphing.

In the earliest grades, students learn how to represent problems
using symbols for unknowns. By 4th grade, they are expected to solve
word problems involving all four operations using letters for unknown
numbers (4.OA.A.2) and to generate and analyze patterns and relationships
(4.OA.C.5). In 5th grade, students begin to use these patterns and relationships to form ordered pairs and to graph these pairs on the coordinate
plane (5.OA.B.3). In 6th and 7th grades, students start to work with ratios,
proportions, expressions, and equations, using these concepts to illustrate
the relationships between quantities in real-world and mathematical sce-
narios. Now, in 8th grade, students need to understand the connections
among proportional relationships, lines, and linear equations as they work
with equations and expressions (8.EE.B).

Mastery of these concepts will help students learn how functions work
as a rule, how to graph a function, and how to compare the properties
of functions. Teachers may find some additional insights offered by the
National Council of Teachers of Mathematics *Principles and Standards for
School Mathematics* (NCTM, 2000) useful as they consider ways to approach
the standards within this cluster, particularly Standard 2 (8.F.A.2), which
focuses on comparing the properties of two functions. NCTM recommends
that teachers prompt students engaging in this work to explain their
observations in their own words, as these explanations may provide a direct
insight into students' thinking. In addition, NCTM encourages teachers to
lead a class discussion about how well the different properties of functions
are illustrated in each of the different representations of functions.

Standard 3 (8.F.A.3), the final standard in Cluster A, asks students to recognize the difference between linear and nonlinear functions. To understand
this difference, they need a solid foundation in exponents, a topic that was
introduced in 5th grade, when students explored the place value system and
learned how to denote powers of 10 (5.NBT.A.2). Students extended their
basic understanding of exponents in 6th grade by writing and evaluating
numerical expressions involving exponents and using the order of operations

on expressions including exponents (6.EE.A.1–2). Here in 8th grade, students move on to work with integer exponents and equations involving exponents.

When planning to address Standard 3, mathematics teachers may want to consult with science teachers, who may be introducing graphs of kinetic energy ($KE = \frac{1}{2}mv^2$), a real-world example of a nonlinear graph. Students tasked with providing examples of such graphs should also be asked to explain not only why the function is nonlinear but also why it is a function.

Use functions to model relationships between quantities

As students become familiar with the concept of functions, they can be expected to begin using functions to model relationships, a skill addressed here, in Cluster B (see Figure 5.4).

8.F.B

> **Figure 5.4 | Use Functions to Model Relationships Between Quantities**
>
> **4.** Construct a function to model a linear relationship between two quantities. Determine the rate of change and initial value of the function from a description of a relationship or from two (*x, y*) values, including reading these from a table or from a graph. Interpret the rate of change and initial value of a linear function in terms of the situation it models, and in terms of its graph or a table of values.
> **5.** Describe qualitatively the functional relationship between two quantities by analyzing a graph (e.g., where the function is increasing or decreasing, linear or nonlinear). Sketch a graph that exhibits the qualitative features of a function that has been described verbally.

While the concept and use of functions will be new to students, the modeling of relationships and the use of graphs that Standard 4 (8.F.B.4) calls for are not. In 5th grade, students were asked to generate patterns using rules and graph ordered pairs on the coordinate plane. In 6th and 7th grades, they spent some time applying ratios and proportional relationships as well as using dependent and independent variables to solve problems using tables, equations, and graphs (6.RP.A.3, 6.EE.C.9, 7.RP.A). These earlier experiences with modeling relationships will help students connect their prior knowledge with the new concept of functions.

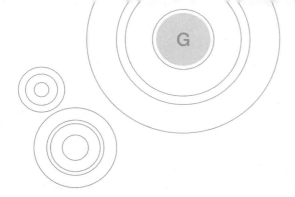

CHAPTER 6

Geometry

The introduction to the high school Geometry standards (CCSSI, 2010c, p. 74) provides valuable insight into the Common Core's approach to this subject matter, particularly its treatment of congruence and transformation. We highly recommend middle school teachers review this material, as the Geometry standards for grades 6–8 provide a critical bridge from content taught in earlier grades to the geometric concepts students will learn in high school.

Figure 6.1 provides an overview of the middle school Geometry (G) domain by grade level.

Taken together, the standards in this domain focus on measurements of geometric figures, congruence and similarity, and problem solving with geometric figures.

Connections to the Standards for Mathematical Practice

As students extend their skills with geometric techniques, they will be implementing multiple mathematical practices.

Using a two-dimensional net as a model to help visualize the surface area of a three-dimensional shape will help students "make sense" of the problem—a skill described in Mathematical Practice Standard 1.

Figure 6.1	**The Geometry Domain: Middle School Overview**	
Grade Level	**Clusters**	**Standards**
Grade 6	**6.G.A** Solve real-world and mathematical problems involving area, surface area, and volume.	6.G.A.1, 6.G.A.2, 6.G.A.3, 6.G.A.4
Grade 7	**7.G.A** Draw, construct and describe geometrical figures and describe the relationships between them.	7.G.A.1, 7.G.A.2, 7.G.A.3
	7.G.B Solve real-life and mathematical problems involving angle measure, area, surface area, and volume.	7.G.B.4, 7.G.B.5, 7.G.B.6
Grade 8	**8.G.A** Understand congruence and similarity using physical models, transparencies, or geometry software.	8.G.A.1, 8.G.A.2, 8.G.A.3, 8.G.A.4, 8.G.A.5
	8.G.B Understand and apply the Pythagorean Theorem.	8.G.B.6, 8.G.B.7, 8.G.B. 8
	8.G.C Solve real-world and mathematical problems involving volume of cylinders, cones and spheres.	8.G.C. 9

As students become comfortable with solving mathematical problems involving volume and surface area, they are expected to begin to use their skills to solve real-world problems, as well. When students describe a problem both as a formula and in context, they are showing evidence of reasoning "abstractly and quantitatively" (Mathematical Practice Standard 2). Using geometric techniques to represent real-world problems also requires students to use skills associated with Mathematical Practice Standard 4, "Model with mathematics."

As they explain how given geometrical criteria (such as the relationship between the circumference and area of a circle) follow logically from geometric principles, or explain a simple geometric proof of the Pythagorean Theorem, they are developing their capacity to "construct viable arguments" (Mathematical Practice Standard 3) and communicate precisely to others (a skill related to Mathematical Practice Standard 6).

Students should have multiple opportunities to "use appropriate tools strategically" (Mathematical Practice Standard 5) throughout the domain, as they draw geometrical figures and transformations using a wide variety of tools. Finally, as students solve problems dealing with composite shapes, solve problems involving missing angles, or transform geometric objects, Mathematical Practice Standard 7 ("Look for and make use of structure") will help them see geometric shapes as combinations of several objects.

Conceptual Pathway Through the Grades

Geometry is a domain that extends from kindergarten through grade 12, although some of the concepts addressed at the middle school level, such as volume and area, are also introduced in the Measurement and Data domain in earlier grades. Figure 6.2 traces the conceptual foundations of the middle school geometry standards throughout elementary school.

Figure 6.2 I **Geometry: Conceptual Pathway to Middle School**	
Grade Level	**Concepts**
Kindergarten	• Basic two- and three-dimensional shapes • Modeling objects in their environment • Construction of more complex shapes
Grade 1	• Composition and decomposition of plane or solid figures • Perspective and orientation of combined shapes • Geometric attributes • How shapes are alike and different
Grade 2	• Shape analysis (using sides and angles) • Decomposition and combination of shapes
Grade 3	• Measurement and estimation of intervals of time, liquid volumes, and masses of objects • The relationship of area to multiplication and to addition • Perimeter as an attribute of plane figures • Difference between linear and area measures • Categories and subcategories of shapes • The partitioning of shapes into parts with equal areas

(continued)

Figure 6.2	**Geometry: Conceptual Pathway to Middle School** *(continued)*	
Grade Level	**Concepts**
Grade 4	• Concepts of angle and how to measure angles • Drawing and identifying lines and angles (right, acute, obtuse) • Classification of shapes by properties of their lines and angles • Lines of symmetry for two-dimensional figures
Grade 5	• Relationship between volume and the operations of multiplication and addition • Graphing points on the coordinate plane • Classifying two-dimensional figures

With this overview complete, it's time to explore the domain's clusters and standards in greater detail.

Grade 6

The standards document identifies the skills needed to solve problems involving area, surface area, and volume as a critical area in 6th grade. This content is also highlighted by the PARCC frameworks document (PARCC, 2011). As noted, critical areas are content identified as meriting substantive treatment, and they should receive priority attention in the classroom.

Solve real-world and mathematical problems involving area, surface area, and volume

There are four standards in the 6th grade Geometry domain, all contained in a single cluster (see Figure 6.3).

The introductory text to the Geometry domain takes care to emphasize that students are expected not simply to use formulas but also to be able to justify the formulas using mathematical models—a methodology intended to ensure students grasp the underlying geometrical concepts. Although it's a departure from the approach taken by many traditional state standards documents, this emphasis on understanding why and

Figure 6.3 | **Solve Real-World and Mathematical Problems Involving Area, Surface Area, and Volume**

1. Find the area of right triangles, other triangles, special quadrilaterals, and polygons by composing into rectangles or decomposing into triangles and other shapes; apply these techniques in the context of solving real-world and mathematical problems.
2. Find the volume of a right rectangular prism with fractional edge lengths by packing it with unit cubes of the appropriate unit fraction edge lengths, and show that the volume is the same as would be found by multiplying the edge lengths of the prism. Apply the formulas $V = l\,w\,h$ and $V = b\,h$ to find volumes of right rectangular prisms with fractional edge lengths in the context of solving real-world and mathematical problems.
3. Draw polygons in the coordinate plane given coordinates for the vertices; use coordinates to find the length of a side joining points with the same first coordinate or the same second coordinate. Apply these techniques in the context of solving real-world and mathematical problems.
4. Represent three-dimensional figures using nets made up of rectangles and triangles, and use the nets to find the surface area of these figures. Apply these techniques in the context of solving real-world and mathematical problems.

how formulas work is not unusual for the Common Core standards. From kindergarten on, the standards explicitly demand that students understand the mathematical concepts they encounter. Whether the concept is operations or geometric formulas, students are expected be able to explain why and how it works.

Here in 6th grade, Standard 1 (6.G.A.1) asks students to calculate the area of given shapes. The 6th grade introductory text (CCSSI, 2010c, p. 40) further stipulates that students be able to discuss, develop, and justify formulas for areas of triangles and parallelograms. They are to use composition and decomposition methods as they create and justify their formulas. As students compose and decompose shapes, they may use paper-and-pencil sketches, geometric models, or dynamic geometry software as aids to visualization. Composition and decomposition methods, addressed throughout the elementary grades, should be familiar to 6th graders. In fact, students first work with these methods in kindergarten, when they are asked to construct more complex shapes from basic ones (K.G.B.6).

The content addressed in Standard 2 (6.G.A.2) is also likely to be familiar to students, as it is very similar to 5.MD.C.5a, which asks them to find the volume of a right rectangular prism with whole-number side lengths using the same methods. Standard 2 simply incorporates the new aspect of fractional edge lengths, representing a small increase in rigor from 5th grade expectations.

Standard 3 (6.G.A.3) is another expansion on prior knowledge. The 5th grade standard 5.G.A.1 requires students to represent real-world and mathematical problems by graphing points in the first coordinate of the coordinate plane. Standard 3 extends this concept to include all four quadrants, connecting the idea with the concepts regarding negative rational numbers found in the Number System domain (see 6.NS.C.7–8, pp. 30–31).

As students begin to explore surface area, some may find it difficult to visualize the unseen faces of three-dimensional shapes. Mastery of Standard 4 (6.G.A.4), calling for the use of nets to represent the surface area of a three-dimensional figure, will assist them as they calculate the surface area of real-world objects and mathematical shapes. This work will draw on students' understanding of three-dimensional objects and on their ability to model objects, concepts developed in earlier grades.

Grade 7

In 7th grade, students build on their understanding of area and use geometry to solve real-life and mathematical problems. The Common Core standards document identifies solving area, surface area, and volume problems, as well as problems involving scale drawings and geometric constructions, as a critical area in 7th grade, and PARCC (2011) also highlights this content in its frameworks document.

At the 7th grade level, the Geometry domain contains six standards, grouped into two clusters.

Draw, construct, and describe geometrical figures and describe the relationships between them

Cluster A addresses how geometrical figures are related to one another (see Figure 6.4).

Figure 6.4 | **Draw, Construct, and Describe Geometrical Figures and Describe the Relationships Between Them**

---

1. Solve problems involving scale drawings of geometric figures, including computing actual lengths and areas from a scale drawing and reproducing a scale drawing at a different scale.
2. Draw (freehand, with ruler and protractor, and with technology) geometric shapes with given conditions. Focus on constructing triangles from three measures of angles or sides, noticing when the conditions determine a unique triangle, more than one triangle, or no triangle.
3. Describe the two-dimensional figures that result from slicing three-dimensional figures, as in plane sections of right rectangular prisms and right rectangular pyramids.

The standards document identifies ratios and proportional reasoning as one of the critical areas in the 6th and 7th grade standards. Standard 1 (7.G.A.1), which deals with solving problems involving scale drawings, mirrors this focus on proportional relationships.

Standard 2 (7.G.A.2) introduces students to informal geometric constructions, asking them to use understandings developed in previous grades (4.G.A, 5.G.B) to help them draw and classify geometric shapes. Teachers may find it beneficial to use dynamic geometry software when teaching this standard, as the technology can help students focus on the results of changing the conditions of the figures rather than on the process of creating the drawings.

Standard 3 (7.G.A.3) requires work with three-dimensional figures, which students must relate to two-dimensional figures using cross sections. The use of geometric models or dynamic geometry software can aid this visualization process. Interpreting or drawing different views of structures may also help students as they develop this skill.

Solve real-life and mathematical problems involving angle measure, area, surface area, and volume

Cluster B extends 7th graders' understanding of area, volume, surface area, and angle measurement to include circles, a variety of three-dimensional shapes, and unknown angles (see Figure 6.5).

Figure 6.5 | **Solve Real-Life and Mathematical Problems Involving Angle Measure, Area, Surface Area, and Volume**

4. Know the formulas for the area and circumference of a circle and use them to solve problems; give an informal derivation of the relationship between the circumference and area of a circle.
5. Use facts about supplementary, complementary, vertical, and adjacent angles in a multi-step problem to write and solve simple equations for an unknown angle in a figure.
6. Solve real-world and mathematical problems involving area, volume, and surface area of two- and three-dimensional objects composed of triangles, quadrilaterals, polygons, cubes, and right prisms.

Students build the understanding necessary to master these concepts throughout the grade levels. In 3rd grade, they are introduced to the concepts of area and liquid volume (3.MD.C). In 4th and 5th grades, they begin to apply the area formula for rectangles to problems (4.MD.A.3) and recognize volume as an attribute of a three-dimensional shape (5.MD.C.3). In 5th and 6th grades, students develop an understanding of and apply the formula for volume, and in 6th grade, they also get their first exposure to the concept of surface area (6.G.A.4).

As Standard 4 (7.G.B.4) shows, by the time students reach 7th grade, they are expected not only to know and use the formulas associated with circles but also to be able to illustrate an informal derivation of the relationship between circumference and area. This emphasis on successfully explaining a formula or relationship rather than just memorizing and using it is a common theme throughout the Common Core standards document— one we have touched on already. Consider that prior to the 7th grade, students calculated the area and perimeter of polygons only. Sensibly, finding the area of a circle is delayed until *after* the focus on ratios and proportions begun in 6th grade and continued in 7th grade enables students to grasp the relationship between circumference and area in a meaningful way.

Standard 5 (7.G.B.5) focuses on angle measurement, introducing the idea of angles formed by intersecting lines. Looking back over the related standards in prior grade levels, we can see that while students learn to classify shapes based on angles very early on, they are not asked to understand angle measurement until 4th grade (4.MD.B, 4.MD.C). That is the point at which they learn what an angle is, how to measure it, and the fact that angle measure is additive. Fourth graders are also asked to solve addition and subtraction problems to find unknown angles on a diagram by using an equation with a symbol for the unknown measure (4.MD.C.7). This is similar to what 7th graders are asked to do here in Standard 5, which differs from 4.MD.C.7 only in the requirements that the problems be multi-step and that students use facts about supplementary, complementary, vertical, and adjacent angles.

Mastery of the standards in the 7th grade Geometry domain will prepare students for 8th grade work on congruence and similarity.

Grade 8

The Common Core standards document identifies an understanding of distance, angle, similarity, congruence, and the Pythagorean Theorem as a critical area in 8th grade. This content is also highlighted by both PARCC (2011) and SBAC (Schoenfeld et al., 2012).

At the 8th grade level, the Geometry domain contains nine standards, grouped into three clusters.

Understand congruence and similarity using physical models, transparencies, or geometry software

The development of geometry in 8th grade begins with the concepts of congruence and similarity (see Figure 6.6).

In 7th grade, students learned to reason about relationships among two-dimensional figures using scale drawings and informal geometric constructions (7.G.A). An understanding of these concepts, along with practice visualizing geometric shapes, prepares students to grasp ideas of congruence and similarity—a critical concept they will continue to explore in high school Geometry (see p. 74 of the Common Core standards document [CCSSI, 2010c] for more detail).

Figure 6.6 | Understand Congruence and Similarity Using Physical Models, Transparencies, or Geometry Software

1. Verify experimentally the properties of rotations, reflections, and translations:
   a. Lines are taken to lines, and line segments to line segments of the same length.
   b. Angles are taken to angles of the same measure.
   c. Parallel lines are taken to parallel lines.
2. Understand that a two-dimensional figure is congruent to another if the second can be obtained from the first by a sequence of rotations, reflections, and translations; given two congruent figures, describe a sequence that exhibits the congruence between them.
3. Describe the effect of dilations, translations, rotations, and reflections on two-dimensional figures using coordinates.
4. Understand that a two-dimensional figure is similar to another if the second can be obtained from the first by a sequence of rotations, reflections, translations, and dilations; given two similar two-dimensional figures, describe a sequence that exhibits the similarity between them.
5. Use informal arguments to establish facts about the angle sum and exterior angle of triangles, about the angles created when parallel lines are cut by a transversal, and the angle–angle criterion for similarity of triangles. *For example, arrange three copies of the same triangle so that the sum of the three angles appears to form a line, and give an argument in terms of transversals why this is so.*

Standard 2 (8.G.A.2) and Standard 4 (8.G.A.4) extend students' understanding of the rigid motions described in Standard 1 (8.G.A.1), which is fundamental to a definition of congruence and similarity transformations.

For an example of a lesson addressing 8.G.A.2 and 8.G.A.3, please see **Sample Lesson 3**.

In 7th grade, students also explored the relationships between angles formed by intersecting lines (7.G.B.5). Now, in 8th grade, they extend that understanding to create informal arguments about a variety of geometric constructions. Although the use of mathematical arguments is a mathematical practice standard, this cluster's Standard 5 (8.G.B.5) is the first content standard to use the term *argument*. Because students in 8th grade may not be familiar with the construction of a mathematical argument, teachers should discuss the elements of a strong mathematical argument, as described in the mathematical practice standards section of the standards document—namely, the use of stated assumptions, definitions, and previously established results; a logical progression of statements; justification and communication of conclusions; and responding to the arguments of others (CCSSI, 2010c, p. 7). Practicing these components will assist students as they move forward into the more formal argumentation processes that will be required in high school mathematics courses.

Understand and apply the Pythagorean Theorem

Cluster B in the 8th grade Geometry domain introduces the Pythagorean Theorem (see Figure 6.7).

Figure 6.7   **Understand and Apply the Pythagorean Theorem**
6. Explain a proof of the Pythagorean Theorem and its converse.
7. Apply the Pythagorean Theorem to determine unknown side lengths in right triangles in real-world and mathematical problems in two and three dimensions.
8. Apply the Pythagorean Theorem to find the distance between two points in a coordinate system.

8.G.B

In order to understand and apply the Pythagorean Theorem and its converse, students must be able to integrate several concepts—exponents, ratios, square roots, and irrational numbers. In 5th grade, students were introduced to exponents as they began to explore the place value system and learn how to denote powers of 10 (5.NBT.A.2). The 6th grade standards extended this understanding of exponents to include numerical expressions involving exponents (6.EE.A.1) and also introduced ratios (6.RP.A). Students in 7th grade expanded their concept of exponents and ratios to include proportions and integer exponents. Another of the 8th grade geometry standards introduces students to square roots (8.EE.A.2), using the square root symbol to represent solutions to equations, and the idea that the square root of 2 is irrational.

In this cluster of the 8th grade Geometry domain, Standards 6, 7, and 8 (8.G.B.6–8) ask students to muster all these understandings. It is somewhat unusual for a standards document to require that students explain a proof of the Pythagorean Theorem, particularly in 8th grade. However, given the emphasis the Common Core standards place on conceptual understanding rather than just rote memorization, the requirement is not entirely surprising.

Solve real-world and mathematical problems involving volume of cylinders, cones, and spheres

Cluster C in the 8th grade Geometry domain addresses problems related to the volume of cones, cylinders, and spheres (see Figure 6.8).

8.G.C

Figure 6.8 | **Solve Real-World and Mathematical Problems Involving Volume of Cylinders, Cones, and Spheres**
9. Know the formulas for the volumes of cones, cylinders, and spheres and use them to solve real-world and mathematical problems.

In 6th grade, students learned to solve real-world problems concerning the volume and surface areas of right rectangular prisms (6.G.A.2); in 7th grade, they worked with problems related to the area of a circle (7.G.B.4). From this foundation, Standard 9 (8.G.C.9) asks 8th graders to expand their skills to include working with three-dimensional shapes that have a circular component, namely, cones, cylinders, and spheres; this work completes students' learning about volume.

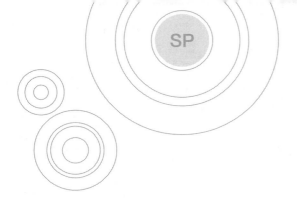

CHAPTER 7

Statistics and Probability

The final domain of the middle school mathematics standards is Statistics and Probability (SP). This domain extends the concepts about data presented in the Measurement and Data domains in grades K–5 and incorporates new concepts about data analysis and chance. While the Common Core Standards Writing Team's *Progressions for the Common Core State Standards in Mathematics* (2011) helps to clarify the content of many domains, we believe its look at the Statistics and Probability domain for grades 6–8 is particularly helpful for all teachers, providing clear and specific examples of the content addressed in the standards.

Figure 7.1 provides an overview of the domain's clusters and standards by grade level.

In this chapter, we'll look at each of these clusters in turn and examine how the standards in each connect and build across clusters, other domains, and grade levels. Let's begin by examining how the standards within Statistics and Probability connect to the Standards for Mathematical Practice.

Connections to the Standards for Mathematical Practice

Students will use several mathematical practice standards as they explore the topic of statistics and probability. As they begin to make sense of statistical variability, they will be asked to summarize and describe distributions in

Figure 7.1	**The Statistics and Probability Domain: Middle School Overview**	
Grade Level	**Clusters**	**Standards**
Grade 6	**6.SP.A** Develop understanding of statistical variability.	6.SP.A.1, 6.SP.A.2, 6.SP.A.3
	6.SP.B Summarize and describe distributions.	6.SP.B.4, 6.SP.B.5
Grade 7	**7.SP.A** Use random sampling to draw inferences about a population.	7.SP.A.1, 7.SP.A.2
	7.SP.B Draw informal comparative inferences about two populations.	7.SP.B.3, 7.SP.B.4
	7.SP.C Investigate chance processes and develop, use, and evaluate probability models.	7.SP.C.5, 7.SP.C.6, 7.SP.C.7
Grade 8	**7.SP.A** Investigate patterns of association in bivariate data.	8.SP.A.1, 8.SP.A.2, 8.SP.A.3, 8.SP.A.4

relation to the context in which data were gathered. This ability to shift viewpoints, that is, to "reason abstractly and quantitatively," is a skill identified in Mathematical Practice Standard 2. Using appropriate tools to illustrate data distributions (Mathematical Practice Standard 5) will help students understand how changes to the data may affect their measures.

When students are asked to write, interpret, or explain mathematical concepts such as statistical variability, they will benefit from the ability to communicate their ideas precisely (Mathematical Practice Standard 6). Students will also use Mathematical Practice Standard 7, "Look for and make use of structure," as they analyze and describe patterns in data sets.

Conceptual Pathway Through the Grades

While the Statistics and Probability domain first appears in 6th grade, the concepts necessary to understand statistical distributions and probability have their roots in previous grades. Figure 7.2 traces these concepts through the elementary grades.

Having looked at the domain and conceptual pathway to mastery as a whole, we will explore the clusters at each grade level in more detail.

Figure 7.2	**Statistics and Probability: Conceptual Pathway to Middle School**
Grade Level	**Concepts**
Kindergarten	• Classification of up to 10 objects
Grade 1	• Organization, representation, and interpretation of data with up to three categories • Asking and answering simple questions about data and categories
Grade 2	• Line plots, picture graphs, and bar graphs with a single unit scale to represent a data set with up to four categories • Solving simple problems using information presented in a bar graph
Grade 3	• Line plots, scaled picture graphs, and scaled bar graphs to represent a data set with several categories • Solving one- and two-step problems using information presented in scaled bar graphs
Grades 4 and 5	• Line plots to display a data set of measurements in fractions of a unit • Solving problems involving operations on fractions by using information presented in line plots

Grade 6

The middle school exploration of this material begins, as expected, by focusing on conceptual understanding. Developing the ability to think statistically is identified as a critical area within the Common Core standards and is content that is highlighted by the *PARCC Model Content Frameworks: Mathematics* (PARCC, 2011).

At the 6th grade level, the Statistics and Probability domain contains five standards, grouped into two clusters.

Develop understanding of statistical variability

The first cluster of the domain's 6th grade standards focuses on statistical variability (see Figure 7.3).

Throughout the elementary grades, the Common Core standards present representing and interpreting data as a supporting area rather than a critical area. This begins to change as students enter middle school, where they are

6.SP.A

| Figure 7.3 | **Develop Understanding of Statistical Variability** |

1. Recognize a statistical question as one that anticipates variability in the data related to the question and accounts for it in the answers. *For example, "How old am I?" is not a statistical question, but "How old are the students in my school?" is a statistical question because one anticipates variability in students' ages.*
2. Understand that a set of data collected to answer a statistical question has a distribution which can be described by its center, spread, and overall shape.
3. Recognize that a measure of center for a numerical data set summarizes all of its values with a single number, while a measure of variation describes how its values vary with a single number.

expected to develop what the standards call "statistical thinking." Statistical thinking includes being able to recognize what a statistical question is and knowing how to describe the parts of a distribution as well as being able to recognize measures of center and measures of variation. Standards 1, 2, and 3 (6.SP.A.1–3) define these critical concepts, and mastering them will help students establish a foundation for thinking statistically. This understanding of statistical concepts is a natural extension from students' work in elementary school, which included drawing line plots and graphs of data sets and using the plots and graphs to solve problems.

Summarize and describe distributions

Mastery of the basic definitions that are the focus of the standards in Cluster A prepares 6th graders to use these concepts to analyze sets of data, the skill covered here in Cluster B (see Figure 7.4).

The Common Core mathematics standards require students to interpret data from various types of graphical representations. While students should be very familiar with displaying data in line plots, picture graphs, and bar graphs from their work in elementary school, Standard 4 (6.SP.B.4) describes some very different types of graphs that will help students answer more complex questions about relationships within a data set. The types of questions that students should be able to summarize are described in Standard 5 (6.SP.B.5).

Figure 7.4 | **Summarize and Describe Distributions**

4. Display numerical data in plots on a number line, including dot plots, histograms, and box plots.
5. Summarize numerical data sets in relation to their context, such as by:
   a. Reporting the number of observations.
   b. Describing the nature of the attribute under investigation, including how it was measured and its units of measurement.
   c. Giving quantitative measures of center (median and/or mean) and variability (interquartile range and/or mean absolute deviation), as well as describing any overall pattern and any striking deviations from the overall pattern with reference to the context in which the data were gathered.
   d. Relating the choice of measures of center and variability to the shape of the data distribution and the context in which the data were gathered.

The description of critical areas in the introduction to the 6th grade mathematics standards is especially illuminating in regard to the content found in Cluster B of the 6th grade Statistics and Probability domain. As students learn about the shape of the data distribution, it is essential they recognize that center can be measured using either the median or the mean and that the choice of measure of center affects the shape of the data distribution. Students should also recognize the usefulness of variability in summarizing data sets, understanding that the differences in variability between data sets can hold as much or more meaning than the measure of center. The patterns that students should be able to recognize include clusters, peaks, gaps, and symmetry within the context of the data.

For an example of a lesson addressing 6.SP.B.4, please see **Sample Lesson 1**.

This understanding of statistical concepts and their work with single data distributions will help students as they begin to compare populations using two data distributions in 7th grade.

Grade 7

In 6th grade, students are introduced to the concept of statistical thinking and using box plots, histograms, and scatter plots to represent data. In 7th grade, they build on these concepts to compare two populations and begin informal work with random sampling. This content is identified as a critical area in the introduction to the 7th grade Common Core standards. The PARCC frameworks document (PARCC, 2011) also highlights the comparison of populations and drawing inferences, while SBAC's *Content Specifications for the Summative Assessment of the Common Core State Standards for Mathematics* (Schoenfeld et al., 2012) highlights probability modeling.

At the 7th grade level, the Statistics and Probability domain contains seven standards, organized into three clusters.

Use random sampling to draw inferences about a population

Cluster A focuses on the ability to use random sampling to draw inferences (see Figure 7.5).

7.SP.A

Figure 7.5	**Use Random Sampling to Draw Inferences About a Population**

1. Understand that statistics can be used to gain information about a population by examining a sample of the population; generalizations about a population from a sample are valid only if the sample is representative of that population. Understand that random sampling tends to produce representative samples and support valid inferences.
2. Use data from a random sample to draw inferences about a population with an unknown characteristic of interest. Generate multiple samples (or simulated samples) of the same size to gauge the variation in estimates or predictions. *For example, estimate the mean word length in a book by randomly sampling words from the book; predict the winner of a school election based on randomly sampled survey data. Gauge how far off the estimate or prediction might be.*

In elementary school and 6th grade, students are expected to create representations of data, but they are not asked to reason about how the data are collected. This changes in 7th grade, with the introduction of probability and random sampling. As described in *Progressions for the Common Core State Standards in Mathematics* (Common Core Standards Writing Team, 2011), because sampling is a concept related to probability, students explore probability (addressed in the third cluster of this domain, 7.SP.C; see pp. 77–78) before beginning work on sampling. Once they have developed an understanding of chance processes and probability models, they are prepared to learn about sampling. Connecting the concepts of statistical modeling and probability will strengthen students' ability to "think statistically," as described in the Common Core introduction to the 6th grade.

Within this cluster, Standard 1 (7.SP.A.1) focuses on developing students' understanding of the use of random sampling, which leads naturally to using that understanding to investigate unknown characteristics of interest about a population, addressed in Standard 2 (7.SP.A.2). Teachers should encourage students to formulate questions and to consider ways to take samples to ensure the samples are representative of a population. In addition to inviting students to generate questions and problems, mathematics teachers may want to collaborate with science teachers on science-related inquiries. This practice allows for a wide variety of questions, illustrating some practical uses of the mathematical content, and ensures consistency of approach across subject areas.

Draw informal comparative inferences about two populations

Cluster B of the 7th grade Statistics and Probability domain builds directly upon Cluster A, in that it asks students to use the data from random samples to draw comparative inferences about two populations (see Figure 7.6).

Figure 7.6 | **Draw Informal Comparative Inferences About Two Populations**   `7.SP.B`

3. Informally assess the degree of visual overlap of two numerical data distributions with similar variabilities, measuring the difference between the centers by expressing it as a multiple of a measure of variability. *For example, the mean height of players on the basketball team is 10 cm greater than the mean height of players on the soccer team, about twice the variability (mean absolute deviation) on either team; on a dot plot, the separation between the two distributions of heights is noticeable.*

4. Use measures of center and measures of variability for numerical data from random samples to draw informal comparative inferences about two populations. *For example, decide whether the words in a chapter of a seventh-grade science book are generally longer than the words in a chapter of a fourth-grade science book.*

Standard 3 (7.SP.B.3) requires students to visually assess the degree of overlap between two data distributions. In elementary school, students compared data categories within a set using picture and bar graphs. In 6th grade, they compared data found in a single data set using data distributions represented in histograms, dot plots, and box plots. Now, in 7th grade, with a firm grasp on the concepts involved in data distributions and sampling, students can begin to compare populations using visual and numerical representations of data. The analysis may require students to use ratio reasoning as they express the difference between the centers (either median or mean) as a multiple of a measure of variability. Reasoning with ratios begins in 6th grade (6.RP.A) and is further developed in the Ratios and Proportional Relationships domain in 7th grade (7.RP.A).

As students begin to draw inferences between two populations, which is the focus of Standard 4 (7.SP.B.4), teachers can support practice with mathematical communication by encouraging students to justify their comparative inferences. To facilitate students' attention to precision, teachers may want to stipulate that students use clear definitions, specify their units of measure (if applicable), cite the specific data and calculated quantities that led them to their inference, and refer to specific relationships in the context of the problem.

Investigate chance processes and develop, use, and evaluate probability models

In Cluster C of the 7th grade Statistics and Probability domain, the focus turns to probability (see Figure 7.7).

The topic of probability is not introduced in the Common Core standards until 7th grade. This is a departure from familiar practice, as many state standards documents introduce basic probability in elementary school. One possible reason for this approach is the emphasis the Common Core places on the connections between statistics and probability. Prior to 7th grade, statistical problem solving focuses on the display of data, with no attention paid to data collection methods. The standards emphasize the need to connect data collection methods with probability,

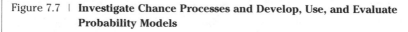

Figure 7.7 | **Investigate Chance Processes and Develop, Use, and Evaluate Probability Models**

5. Understand that the probability of a chance event is a number between 0 and 1 that expresses the likelihood of the event occurring. Larger numbers indicate greater likelihood. A probability near 0 indicates an unlikely event, a probability around 1/2 indicates an event that is neither unlikely nor likely, and a probability near 1 indicates a likely event.

6. Approximate the probability of a chance event by collecting data on the chance process that produces it and observing its long-run relative frequency, and predict the approximate relative frequency given the probability. *For example, when rolling a number cube 600 times, predict that a 3 or 6 would be rolled roughly 200 times, but probably not exactly 200 times.*

7. Develop a probability model and use it to find probabilities of events. Compare probabilities from a model to observed frequencies; if the agreement is not good, explain possible sources of the discrepancy.

   a. Develop a uniform probability model by assigning equal probability to all outcomes, and use the model to determine probabilities of events. *For example, if a student is selected at random from a class, find the probability that Jane will be selected and the probability that a girl will be selected.*

   b. Develop a probability model (which may not be uniform) by observing frequencies in data generated from a chance process. *For example, find the approximate probability that a spinning penny will land heads up or that a tossed paper cup will land open-end down. Do the outcomes for the spinning penny appear to be equally likely based on the observed frequencies?*

8. Find probabilities of compound events using organized lists, tables, tree diagrams, and simulation.

   a. Understand that, just as with simple events, the probability of a compound event is the fraction of outcomes in the sample space for which the compound event occurs.

   b. Represent sample spaces for compound events using methods such as organized lists, tables, and tree diagrams. For an event described in everyday language (e.g., "rolling double sixes"), identify the outcomes in the sample space which compose the event.

   c. Design and use a simulation to generate frequencies for compound events. *For example, use random digits as a simulation tool to approximate the answer to the question: If 40% of donors have type A blood, what is the probability that it will take at least 4 donors to find one with type A blood?*

so the two topics are introduced together rather than spread out over earlier grades. In addition, students' understanding of probability is enhanced when they have a firm foundation in ratios, a topic first introduced in 6th grade.

As is often the case in the Common Core standards, students are expected to develop a conceptual understanding before beginning to apply that understanding. Here in Cluster C of the 7th grade Statistics and Probability domain, Standard 5 (7.SP.C.5) asks students to understand how the likelihood of chance events is expressed, and Standard 6 (7.SP.C.6) asks them to use that understanding to determine the probability of a chance event. As students begin to determine probabilities of events, teachers can model the use of terminology associated with probability, including *probability models, compound events,* and *chance processes.* Students will benefit from an understanding of fractions, ratios, and percentages as they calculate the probabilities of events. This understanding of fractions stems from their work in elementary school with fractional models and operational algorithms, their 6th grade work with ratios, and the introduction to proportions they received in 7th grade.

In *Progressions for the Common Core State Standards in Mathematics* (2011), the Common Core Standards Writing Team explains that in 7th grade, students must make the connections between relative frequency and probability. That is, if the ratios of categories within a sample are known (e.g., there are three red socks and seven blue socks in a drawer), then students can predict the frequency—the probability—of a given event (e.g., drawing a red sock from the drawer). If the ratio of categories is unknown, random sampling can lead to an understanding of what that ratio is. This understanding can help students to gain a deeper insight into both processes.

As mentioned in the descriptions of Clusters A and B of this domain, mathematics teachers teaching 7.SP.C.5 and 7.SP.C.6 may benefit from collaborating with science teachers regarding probability and statistics. Often, middle school science curricula include these concepts as well.

The skills related to probability, random sampling, and drawing comparative inferences will be further developed in 8th grade.

Grade 8

At the 8th grade level, the Statistics and Probability domain contains four standards in a single cluster.

Investigate patterns of association in bivariate data

The statistics and probability standards in this cluster (see Figure 7.8) introduce the concept of bivariate data, which the writers of the Common Core consider to be a supporting concept rather than a critical one. Although Standards 1, 2, 3, and 4 (8.SP.A.1–4) should not be a primary focus of 8th grade mathematics instruction, the content still needs to be covered, as it is important knowledge that will assist students as they enter high school.

Figure 7.8 | **Investigate Patterns of Association in Bivariate Data**

8.SP.A

1. Construct and interpret scatter plots for bivariate measurement data to investigate patterns of association between two quantities. Describe patterns such as clustering, outliers, positive or negative association, linear association, and nonlinear association.
2. Know that straight lines are widely used to model relationships between two quantitative variables. For scatter plots that suggest a linear association, informally fit a straight line, and informally assess the model fit by judging the closeness of the data points to the line.
3. Use the equation of a linear model to solve problems in the context of bivariate measurement data, interpreting the slope and intercept. *For example, in a linear model for a biology experiment, interpret a slope of 1.5 cm/hr as meaning that an additional hour of sunlight each day is associated with an additional 1.5 cm in mature plant height.*
4. Understand that patterns of association can also be seen in bivariate categorical data by displaying frequencies and relative frequencies in a two-way table. Construct and interpret a two-way table summarizing data on two categorical variables collected from the same subjects. Use relative frequencies calculated for rows or columns to describe possible association between the two variables. *For example, collect data from students in your class on whether or not they have a curfew on school nights and whether or not they have assigned chores at home. Is there evidence that those who have a curfew also tend to have chores?*

Students have learned about statistical concepts throughout their schooling, beginning with simple graphs and plots in elementary school. This elementary understanding is developed further in middle school, with the introduction of statistical variability and data distributions in 6th grade and data collection and comparative inferences in 7th grade. Now, in 8th grade, students also begin to use linear equations to represent and solve problems in the coordinate plane (see 8.EE.B, 8.EE.C.8). The foundations these skills provide allow students to master the construction of scatter plots and lines of best fit for bivariate measurement data. Students will benefit from a firm understanding of ratios and proportions as they interpret data, which they focused on primarily in the 6th and 7th grades and go on to connect with linear equations in 8th grade (8.EE.B.5).

Students' work with describing univariate measures and developing inferences in 6th and 7th grades will allow them to describe the patterns identified in Standard 1 (8.SP.A.1): clustering, outliers, positive and negative associations, and nonlinear associations. What students need to describe for bivariate data is similar to what students needed to describe according to the 6th grade standards: quantitative measures of center, variability, overall pattern, and striking deviations from that pattern.

Teachers should encourage 8th graders to formulate questions that address patterns of association between two quantities. Again, teachers may also wish to collaborate with science teachers on science-related inquiries. The use of questions about relevant science topics illustrates practical uses of mathematical content and ensures a consistency of approach across subject areas.

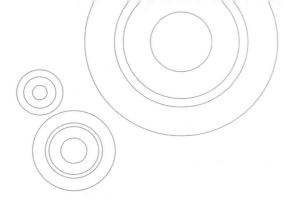

CHAPTER 8

Guidance for Instructional Planning

In this chapter, we provide a brief tutorial on designing lesson plans using the types of instructional strategies that appear in this guide's sample lessons. It includes a step-by-step outline for the development of lessons that make best use of proven instructional strategies and will help you ensure students master the new and challenging content represented by the Common Core standards.

The Framework for Instructional Planning

To identify and use effective strategies to develop these lessons, we draw on the instructional planning framework developed for *Classroom Instruction That Works, 2nd edition* (Dean et al., 2012), presented in Figure 8.1.

The Framework organizes nine categories of research-based strategies for increasing student achievement into three components. These components focus on three key aspects of teaching and learning: creating an environment for learning, helping students develop understanding, and helping students extend and apply knowledge. Let's take a close look at each.

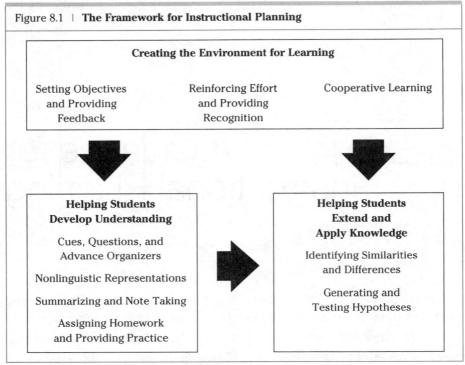

Figure 8.1 | **The Framework for Instructional Planning**

Creating the Environment for Learning

Setting Objectives and Providing Feedback

Reinforcing Effort and Providing Recognition

Cooperative Learning

Helping Students Develop Understanding

Cues, Questions, and Advance Organizers

Nonlinguistic Representations

Summarizing and Note Taking

Assigning Homework and Providing Practice

Helping Students Extend and Apply Knowledge

Identifying Similarities and Differences

Generating and Testing Hypotheses

Source: From *Classroom Instruction That Works, 2nd ed.* (p. xvi) by Ceri Dean, Elizabeth Hubbell, Howard Pitler, and Bj Stone, 2012, Alexandria, VA: ASCD; and Denver, CO: McREL. Copyright 2012 by McREL. Adapted with permission.

Creating the environment for learning

Teachers create a positive environment for learning when they ensure that students are motivated and focused, know what's expected of them, and regularly receive feedback on their progress. When the environment is right, students are actively engaged in their learning and have multiple opportunities to share and discuss ideas with their peers.

A number of instructional strategies that help create a positive environment for learning may be incorporated into the lesson design itself. Other aspects, such as reinforcing effort and providing recognition, may not be a formal part of the lesson plan but are equally important. The following strategies are essential for creating a positive environment for learning:

- Setting objectives and providing feedback
- Reinforcing effort and providing recognition
- Cooperative learning

Helping students develop understanding

This component of the Framework focuses on strategies that are designed to help students work with what they already know and help them integrate new content with their prior understanding. To ensure that students study effectively outside class, teachers also need strategies that support constructive approaches to assigning homework. The strategies that help students develop understanding include the following:

- Cues, questions, and advance organizers
- Nonlinguistic representations
- Summarizing and note taking
- Assigning homework and providing practice

Helping students extend and apply knowledge

In this component of the Framework, teachers use strategies that prompt students to move beyond the "right answers," engage in more complex reasoning, and consider the real-world connections and applications of targeted content and skills, all of which help students gain flexibility when it comes to using what they have learned. The following strategies help students extend and apply knowledge:

- Identifying similarities and differences
- Generating and testing hypotheses

Figure 8.2 illustrates the three major components of teaching and learning described in *Classroom Instruction That Works*, along with the nine types, or categories, of strategies that further define the components and point you toward activities that will suit your learning objectives and support your students' success.

Figure 8.2 | **Framework Components and the Associated Categories of Instructional Strategies**

Component	Category	Definition
Creating the Environment for Learning	Setting Objectives and Providing Feedback	Provide students with a direction for learning and with information about how well they are performing relative to a particular learning objective so they can improve their performance.
	Reinforcing Effort and Providing Recognition	Enhance students' understanding of the relationship between effort and achievement by addressing students' attitudes and beliefs about learning. Provide students with non-material tokens of recognition or praise for their accomplishments related to the attainment of a goal.
	Cooperative Learning	Provide students with opportunities to interact with one another in ways that enhance their learning.
Helping Students Develop Understanding	Cues, Questions, and Advance Organizers	Enhance students' ability to retrieve, use, and organize what they already know about a topic.
	Nonlinguistic Representations • Graphic Organizers • Pictures and Pictographs • Mental Images • Kinesthetic Movement • Models/Manipulatives	Enhance students' ability to represent and elaborate on knowledge using mental images.
	Summarizing and Note Taking	Enhance students' ability to synthesize information and organize it in a way that captures the main ideas and supporting details.
	Providing Practice, and Assigning Homework	Extend the learning opportunities for students to practice, review, and apply knowledge. Enhance students' ability to reach the expected level of proficiency for a skill or process.

Component	Category	Definition
Helping Students Extend and Apply Knowledge	Identifying Similarities and Differences • Comparing • Classifying • Creating/Using Metaphors • Creating/Using Analogies	Enhance students' understanding of and ability to use knowledge by engaging them in mental processes that involve identifying ways in which items are alike and different.
	Generating and Testing Hypotheses	Enhance students' understanding of and ability to use knowledge by engaging them in mental processes that involve making and testing hypotheses.

Figure 8.2 | **Framework Components and the Associated Categories of Instructional Strategies** *(continued)*

Source: From *Classroom Instruction That Works, 2nd ed.* (p. xviii) by Ceri Dean, Elizabeth Hubbell, Howard Pitler, and Bj Stone, 2012, Alexandria, VA: ASCD; and Denver, CO: McREL. Copyright 2012 by McREL. Adapted with permission.

Lesson Development, Step by Step

To help you get started developing lessons that incorporate these strategies, we provide a step-by-step process to ensure that you've had an opportunity to consider where within a lesson the various strategies might be used most effectively. Those steps are as follows:

1. Identify the focus for the lesson.
2. Determine how learning will be assessed.
3. Determine the activities that will start the lesson.
4. Determine the activities that will engage students in learning the content.
5. Determine the activities that will close the lesson.

Let's look now at the details of each step and how you might incorporate the nine effective instructional strategies associated with the Framework's three components. We'll reference the sample lessons in this guide to illustrate particular aspects of this approach.

Step 1: Identify the focus for the lesson

The critical first step in crafting a lesson is to identify what students should learn as a result of their engagement in the lesson activities. Setting objectives for students also means establishing the guidelines for your development of the lesson: namely, that you will select and develop only those activities that will help students meet the objectives set. A learning objective is built directly from a standard; the objectives found in this guide's sample lessons are constructed from Common Core standards and listed under the heading "Common Core State Standards—Knowledge and Skills to Be Addressed."

Clarifying learning objectives. To ensure that students are clear about what they will learn, you will want your lesson plans to include more specific statements of the objectives in clear, student-friendly language. Some teachers accomplish this by using stems such as "I can . . ." or "We will be able to . . ." or "Students will be able to . . ." and providing a paraphrased version of the standard, simplifying the language as necessary. In the sample lessons for this guide, such specifics may be found under the headings "Knowledge/Vocabulary Objectives" and "Skill/Process Objectives" and prefaced by either "Students will understand . . ." or "Students will be able to. . . ."

Identifying essential questions and learning objectives. Framing the lesson's objectives under a broader essential question provides students with alternate avenues to find personal relevance and can energize them to seek answers as they begin a unit or lesson. The essential question properly focuses on the broader purpose of learning, and it is most effective when it is open-ended and not a question that can be easily answered. Each of the sample lessons includes an essential question—the learning objectives reframed to clarify for students what value the lesson holds for them.

Identifying foundational knowledge and possible misconceptions related to the learning objectives. As you develop learning objectives for a lesson, consider the other skills students will need to use but that will not be the explicit focus of instruction or assessment. Our discussions of each standard in this guide identify the critical knowledge and skills that students are assumed to have mastered or practiced in lessons prior to learning the new content. In the sample lessons, you'll find these standards under the heading "Common Core State Standards—Prior Knowledge and Skills to Be Applied."

Step 2: Determine how learning will be assessed

As important as identifying the learning objective for a lesson is identifying the criteria you will use to determine if students have met that objective. You will want to be clear about the rigor identified in the Common Core standards. As you develop scoring tools, such as checklists and rubrics that define the various levels of performance related to the objective's knowledge or skill, it is important to review the details of the objective's underlying standard to be sure you are looking for the appropriate level of mastery.

Assessing prior knowledge. Step 2 involves planning how to measure students' prior knowledge, especially the knowledge identified in Step 1 as prerequisite to mastery of the learning objective. For example, you might ask students to complete a short problem or share reflections on their prior experiences with similar tasks. This approach may also surface any lingering student misconceptions that you'll want to address before proceeding.

Providing feedback. This part of the planning process also includes deciding how you will provide students with feedback on their progress toward the outcome. Providing feedback is an important aspect of creating the environment for learning because understanding what good performance looks like, how to judge their own performance relative to a benchmark, and what they need to do to improve their performance helps students develop a sense of control over their learning. During lesson planning, you might also consider how peers can give their classmates feedback on progress toward the stated objective.

Step 3: Determine the activities that will start the lesson

Step 3 of the planning process concerns the sequence of activities at the start of the lesson, which relate to the "Creating the Environment for Learning" component of the Framework for Instructional Planning. The beginning of each lesson should be orchestrated to capture students' interest, communicate the learning objectives, and encourage their commitment to effort.

Communicating learning objectives. You can share learning objectives by stating them orally, but be sure to post them in writing for reference throughout the lesson. Doing so not only reminds the class of the objectives' importance but also ensures that even students who weren't paying close attention or who came in late can identify what they are working to achieve.

Identifying the essential question and providing a context. Students engage in learning more readily when they can see how it connects to their own interests. The essential question you provide at the beginning of the lesson helps orient them to the purpose for learning. Students will also have a greater sense of involvement if you share with them what activities they'll be engaged in and how these activities will help build their understanding and skill. The sample lessons in this guide present this preview under the heading "Activity Description to Share with Students." It is something you might read aloud or post, along with the objectives and essential questions, as you create the environment for learning at the beginning of a lesson. To encourage greater involvement, you might also ask students to set personal goals based on the learning objectives in each activity. These personal goals may translate the learning objective to immediate goals that resonate for each student.

Reinforcing effort. As you develop the activities for the lesson, look for natural points where you might build in opportunities for students to receive the encouragement they need to continue their work. To reinforce student effort, we need to help students understand the direct connection between how hard they work and what they achieve. It's another way in which teachers can provide students with a greater sense of control over their own learning.

Step 4: Determine the activities that will engage students in learning the content

At Step 4 we are at the crux of the lesson, deciding what students will do to acquire, extend, and apply knowledge or skills. This stage of planning includes identifying when formative assessment should take place, when you will provide students feedback from the assessment, and how you will ensure that students have a clear understanding of how they are doing. And, of course, you will need to decide which instructional activities will best serve the lesson's primary purposes, considering whether the activities need to focus on helping students acquire new knowledge and skill or extend and refine what they've already learned.

Choosing activities and strategies that develop student understanding. When your aim is to help students understand new information or a new process, then you will want to design activities that incorporate strategies associated with that component of the Framework for Instructional Planning. These are the strategies that help students access prior knowledge and organize new learning. Students come to every lesson with some prior knowledge, and the effective use of strategies such as using cues, questions, and advance organizers can enhance students' ability to retrieve and use what they already know about a topic in order to access something new. You can help students access and leverage their prior knowledge through simple discussion, by providing "KWL"-type advance organizers, by having students read or listen to short texts related to the targeted content, or any of a number of ways. Activities incorporating the use of nonlinguistic representations (including visualization) in which students elaborate on knowledge, skills, and processes are other good ways to help students integrate new learning into existing knowledge. The strategies of note taking and summarizing also support students' efforts to synthesize information through the act of organizing it in a way that captures its main ideas and supporting details or highlights key aspects of new processes. Finally, homework can help students learn or review new content and practice skills so that they can more quickly reach the expected level of proficiency. However, you will want to think carefully about your homework practices, as the research on what makes homework effective shows mixed results. Dean and colleagues (2012) recommend that teachers design homework assignments that directly support learning objectives. Students need to understand how homework serves lesson objectives, and once homework is completed, it is important that teachers provide feedback on the assignment.

Choosing activities and strategies that help students extend and apply knowledge. When your aim is to help students extend or apply their knowledge or master skills and processes, they will need opportunities to practice independently. What are beneficial are activities that involve making comparisons, classifying, and creating or using metaphors and analogies. Research summarized in the second edition of *Classroom Instruction That Works* indicates that these strategies, associated with the "Helping Students Extend and Apply Knowledge" component of the Framework for Instructional Planning, are a worthwhile use of instructional time. They help to raise students' levels of understanding and improve their ability to use what they learn. Because students need to understand the concepts or skills that they're comparing, you are more likely to insert these activities later in a lesson than at the outset.

Remember, too, that strategies that help students generate and test hypotheses are not meant just for science classrooms. They are a way to deepen students' knowledge by requiring them to use critical-thinking skills, such as analysis and evaluation.

Grouping students for activities. Cooperative learning can be tremendously beneficial, whether students are developing a new skill or understanding or applying or extending it. With every lesson you design, consider when it makes sense to use this strategy, what kind of student grouping will be most beneficial, and how these groups should be composed. Cooperative learning is a strong option, for example, when you want to differentiate an activity based on student readiness, interest, or learning style. Consider, too, that students' learning experiences will be different depending on whether you permit them to self-select into groups of their choosing or assign their group partners, whether the groups are larger (four or five students) or smaller (e.g., pair work), and whether these groups are homogeneous or heterogeneous.

Providing students with the opportunity to share and discuss their ideas with one another in varying cooperative learning arrangements lays a foundation for the world beyond school, which depends on people working interdependently to solve problems and to innovate. Interacting with one another also deepens students' knowledge of the concepts they are learning; in other words, talking about ideas and listening to others' ideas helps students understand a topic and retain what they've learned, and it may send their thinking in interesting new directions.

Step 5: Determine the activities that will close the lesson

Bringing the lesson to a close provides an opportunity for you and students to look back on and sum up the learning experience.

During this part of the lesson, you want to return to the learning objectives and confirm that you have addressed each of them. This can be approached in one or more ways—through informal sharing, formative assessment, or even summative assessment. Students benefit from the opportunity to gauge their progress in learning. You might prompt them to reflect on the lesson in a journal entry, learning log, or response card, which can easily serve as an informal check for understanding. Note that asking students to share what they found most difficult as well as what worked well can provide you with insight you can apply during the next lesson and can use to refine the lesson just completed.

Depending upon the nature of the objective and whether the lesson appears late in the unit, you may elect to conduct a formal summative assessment. Alternatively, you may identify a homework assignment tied to the learning objective, making sure that students understand how the assignment will help them deepen their understanding or develop their skill.

* * *

In the remaining pages of this guide, we offer sample lesson plans based on the Common Core State Standards for Mathematics, the Framework for Instructional Planning, and the steps just outlined.

Understanding Statistical Questions and Describing Data

Course: 6th grade Mathematics
Length of Lesson: Two hours; two 60-minute class periods

Introduction

In the Common Core State Standards (CCSSI, 2010c), students in kindergarten through grade 5 represent and interpret data as part of the Measurement and Data domain. In grade 6, students begin a formal approach to statistics, with an emphasis on understanding the purpose of collecting and analyzing data. They are asked to identify statistical questions and analyze data by center, spread, and shape. This is a conceptual approach that has not typically been used at the 6th grade level, and it's an example of how the transition to the Common Core may require teachers to create new lessons or adapt existing ones.

This lesson emphasizes critical vocabulary and the process of analyzing statistical data. The first class period focuses on developing both an understanding of the components of a statistical question and essential vocabulary. The second class period focuses on describing data by center, spread, and overall shape. To help refine and cement what students learn in this lesson, they will need subsequent problems and practice.

Strategies from the Instructional Planning Framework

- *Creating the Environment for Learning:* The essential question ("How are statistical data described?") and learning objective ("To understand the components of a statistical question and to describe statistical data using appropriate terminology") provide focus for the lesson. Discussing these lesson components and clarifying the objectives in student-friendly language are ways to help students understand what is expected of them. Providing feedback also plays an important role in this lesson. Students receive feedback from the teacher and other students as they craft statistical questions and as they work with new vocabulary. This helps students self-correct and allows them to make revisions during the learning.

  Cooperative learning occurs during informal partnering, and all students have the opportunity to contribute to discussions and learn from one another. A relevant context is provided to help students understand how the concepts of this lesson are reflected in situations beyond the mathematics classroom.

- *Helping Students Develop Understanding:* Guided practice, which includes an organized format for note taking, is used to help students develop their understanding of the lesson's concepts. Vocabulary instruction is explicit and incorporates nonlinguistic representations to help students build mental images. To support the synthesis of lesson information, students are asked to summarize learning during the lesson closure.

- *Helping Students Extend and Apply Knowledge:* Students are asked to apply what they are learning to predict the center, spread, and shape of data for a statistical question that they generate. They are also asked to provide reasons for their predictions using newly acquired statistical language.

Common Core State Standards—Knowledge and Skills to Be Addressed

Standards for Mathematical Practice

MP3 Construct viable arguments and critique the reasoning of others.

MP6 Attend to precision.

Standards for Mathematical Content

Domain: Statistics and Probability

Cluster: Develop Understanding of Statistical Variability

6.SP.A.1 Recognize a statistical question as one that anticipates variability in the data related to the question and accounts for its answers. *For example, "How old am I?" is not a statistical question, but "How old are the students in my school?" is a statistical question because one anticipates variability in students' ages.*

6.SP.A.2 Understand that a set of data collected to answer a statistical question has a distribution which can be described by its center, spread, and overall shape.

6.SP.A.3 Recognize that a measure of center for a numerical data set summarizes all of its values with a single number, while a measure of variation describes how its values vary with a single number.

Common Core State Standards—Prior Knowledge and Skills to Be Applied

Although grade 6 marks students' first formal study of statistics, they have been collecting and displaying data in line plots and bar graphs since 1st grade. These early learning experiences are not directly aligned with the objectives in this lesson, but they do provide a foundation for students' new understanding. Teachers may reference the standards in the cluster "Represent and interpret data" throughout the elementary-level Measurement and Data domain (1.MD.C, 2.MD.D, 3.MD.B, 4.MD.B, 5.MD.B) to gain additional insight on these early learning experiences.

Teacher's Lesson Summary

In this lesson, students learn to recognize statistical questions and describe data by center, spread, and shape using graphical displays. They build an understanding of essential vocabulary and begin to analyze data. It's essential to help students make connections between the statistics concepts in the lesson and day-to-day applications of data analysis. Students will be grouped in pairs for collaboration and sense-making activities during most of the lesson. "Clock partners," or a similar structure for determining student pairs, is appropriate for this lesson and helps encourage a positive learning environment. Student pairs should be created based on their willingness to cooperate, collaborate, and provide descriptive feedback to each other.

Essential Question: How are statistical data described?

Learning Objective: To understand the components of a statistical question and to describe statistical data using appropriate terminology.

Knowledge/Vocabulary Objectives

At the conclusion of this lesson, students will

- Understand the components of a statistical question.
- Understand that data can be described by center, spread, and shape.
- Understand that measures of center and measures of variability are used to describe data.
- Understand and be able to explain the meaning of the following terms: *statistical question; center, spread, and shape of data; histogram; box plot, mean, median, range, measure of center;* and *measure of variability.* (Interquartile range may be discussed, but it is not the focus for this lesson.)

Skill/Process Objectives

At the conclusion of this lesson, students will be able to

- Justify and communicate reasoning as to why a question is a statistical question.
- Predict the center, spread, and shape of the data resulting from a statistical question.

Resources/Preparation Needed

- A prepared handout covering a discussion protocol for writing, discussing, and revising statistical questions (see Figure A), one per student pair
- Prepared sets of vocabulary cards addressing center, spread, and shape (see Figure B), precut, shuffled, and placed in plastic bags for easy distribution, one per student pair
- A prepared guided-practice handout for note taking and sense making (see Figure C), one per student

Activity Description to Share with Students

It's an everyday occurrence for people to collect data for the purpose of answering questions. If you are curious about what kind of athletic shoe 6th graders are most likely to wear, you could collect data to answer that question. If you wonder how many pets, on average, the students in our school have, you might collect data to answer that question. Questions are also a way to get the information you need to make good decisions. For example, suppose your family wants to adopt a pet, but you're uncertain what type of pet will be the best fit for your busy lives. Or suppose your family is trying to determine which vacation location will be fun for everyone yet still within the family budget. Collecting data on these questions could help you make informed—and, therefore, *better*—decisions.

This lesson will teach you a process for getting answers from data. It will help you understand how to develop statistical questions and how to analyze the data those questions generate. These skills will not only help you find answers to all kinds of questions but also help you make more informed decisions in your daily life.

Lesson Activity Sequence—Class #1

Start the Lesson

Post and lead a quick discussion of the lesson's essential question and objective. Ask students to turn to the person they're sitting next to and discuss the following questions:

- What does the term *statistics* mean to you?
- When have you or a member of your family used statistics to help make a decision?

After two or three minutes of discussion, bring the whole group back together and ask a few students to share examples. It is recommended that you make note of the general level of background knowledge, of any general misconceptions that surface, and of students who seem particularly engaged or disengaged with the topic. This will help you guide the discussion more effectively and plan more appropriately for future learning experiences.

Engage Students in Learning the Content

1. Explain to students that statistics is a type of mathematics used to collect, organize, and interpret data and that the process of collecting data begins with a good statistical question. Statistical questions are written in such a way to expect variance (differences), and it is these differences that make them interesting to study. Provide the following questions to student pairs, and ask them to discuss and decide whether or not the question is a statistical question. Remind them to jot down their reasoning.

   A Statistical Question or NOT a Statistical Question?

   - How old am I?
   - How old are the students in my school?
   - Do I buy or pack my lunch?
   - Do the students in my school buy or pack their lunch?
   - How long did I wait in line at McDonald's last night?
   - How long do customers wait in line at McDonald's?

   Allow approximately two or three minutes for the pairs to agree on the classification of these questions, and then call the class together again for a debrief. Go through the questions one by one, helping students understand the three key components of a statistical question and capturing this information on a chart as you explain it:

   The Components of a Statistical Question

   - It targets a specific audience (e.g., grade 6 students, students who play soccer, students who travel to school by bus, etc.).
   - It expects variance (differences) in responses.
   - It can be answered with data.

2. Distribute the **Discussion Protocol for Statistical Questions** (see **Figure A**, p. 104), a handout that walks student pairs through the process of coming up with two statistical questions; justifying these questions in writing; discussing these questions and defending their reasoning orally; providing feedback on the others' statistical questions and reasoning; and making revisions to their questions, based on peer feedback. Circulate throughout the room during this

activity, answering questions and clarifying misconceptions as needed. Then, wrap up this introduction to statistical questions with a whole-class discussion, asking a volunteer from each pair to share one of their statistical questions and identify the components that make it so. Mention to students that they will need their question sets during the next day's lesson, when they'll use them to predict the center, spread, and shape of data.

3. Move on to the vocabulary development activity. Distribute a packet of the prepared **Vocabulary Sort Cards** (see **Figure B,** pp. 105–106) to each student pair. Explain to students that this activity will build their understanding of essential vocabulary and help prepare them for the second part of the lesson, during which they will use the vocabulary terms to describe data. Ask the student pairs to match each Word Card with the correct Description Card and Example Card. When pairs have finished the assignment (three to five minutes), give them another five minutes to compare their matches with those of another pair of students. Encourage students, if they note differences in the way their classmates have matched their cards, to discuss why they made the choices that they made.

4. Back in the whole-class group, review each term to help students make a personal connection with the word and encourage them to take notes. Questions to ask might include

- Where have you encountered this word before?
- What are examples of how you have seen this term used in mathematics?
- Does this word have meaning in contexts other than mathematics?
- What might help you remember this word?

Tell students that data can be described by *center, spread,* and *shape* and that the vocabulary terms in the card sets (*statistics, statistical question, mean, median, range, skewed distribution, histogram,* and *box plot*) will help them describe data. Ask students to arrange the terms into four groups:

- Terms that describe center
- Terms that describe spread
- Terms that describe shape
- Other terms

Use the examples of the histogram and box plot depicted on the cards to initiate a discussion about the center, spread, and shape of the graphs. Draw attention to the words *measure of center* and *measure of variability* in the description of the terms *mean, median,* and *range.* Help students understand that a measure of center for a numerical data set summarizes all of its values with a single number, while a measure of variation describes how its values vary with a single number.

Close the Lesson

Ask students to work with their partner to answer the following questions, recording answers in their mathematics notebooks:

- What are components of a statistical question?
- What are ways that data can be described?
- What are two types of graphical displays used in today's lesson?
- What are measures of center and measures of variability?

Allow approximately five minutes for students to complete the questions, and then prompt the pairs to share their answers aloud. This closing activity provides an opportunity for informal formative assessment, so it is important to listen carefully to students' responses and clarify understanding when misconceptions come to light.

Lesson Activity Sequence—Class #2

Start the Lesson

Direct students to the essential question and the learning objective posted from the previous day, and tell them that today's lesson will focus on describing data by center, spread, and shape. Display the vocabulary words from the card sort and review their descriptions.

Engage Students in Learning the Content

1. Distribute the **Describing Data Guided Practice Handout** (**Figure C,** pp. 107–108). Students have had little or no experience describing data, so it is important to guide students through the process and provide descriptive feedback to the statements they offer. Display both Spelling Scores histograms, and ask student pairs to discuss how they might describe the *shape* of the data. Allow one to two

minutes for discussion, and then ask a few students to share their descriptions aloud. Help students craft appropriate descriptions, and make sure all students record these on their handout. Continue in this same way with *spread* and *center*. (Students typically feel more confident describing the data's shape and spread, but center may be a difficult concept as they begin this work. Using a sentence starter such as "Pre-test spelling scores have a _____ distribution and are grouped around _____," may help students begin this task.)

Describing Data: Questions About the Histogram

- How might you describe the shape of the first histogram?
- How does the histogram from the pre-test compare to that of the post-test?
- How does the measure of center change from the first graph to the second?
- How might you describe the spread of data in the histograms?

Here are some illustrative sample statements describing the histograms:

- *Shape:* Pre-test spelling scores have a skewed distribution and are grouped around the lowest scores.
- *Center and Spread:* Pre-test spelling scores have a wider, or larger, variance. Since more students had initial low scores, the center is lower than for the post-test.
- *Shape:* Post-test spelling scores have a normal shape. This means that the data are symmetrical with about the same number of data points on each side of center. This is sometimes called a bell shape.
- *Center and Spread:* Post-test spelling scores are clustered together with a narrower variance than the pre-test scores and have a higher center. This means that students' scores on the post-test were higher than their scores on the pre-test.

2. After students have recorded descriptive statements about the histograms, proceed to description of the box plots. Use questions similar to the histogram questions to help students develop descriptive statements for the center, spread, and shape of the box plots.

Sample statements describing the box plots of Minneapolis:

- The shape is close to normal. That means that the data are almost symmetrical, with about the same number of data points on both sides of the middle.

- The high and low temperatures are a similar distance from the cluster (box).
- There is a wider range in the cluster (box) than for the Honolulu cluster.
- The median (middle number) lies close to the center of the cluster (box).

Sample statements describing the box plots of Honolulu:

- The shape is skewed to the higher temperatures.
- The range of temperatures is not as wide as for the Minneapolis plot.
- The low temperature is a greater distance from the cluster (box) than the high temperature.
- The median (middle number) lies at the lower end of the cluster (box).

3. Next, provide student pairs with a question (see the suggestions listed or use a student-generated question) and ask them to predict the center, spread, and shape of the data. Encourage them to use appropriate terminology. It may be helpful to provide a list of vocabulary terms or sentence starters for student reference. Repeat this process with other questions as time allows.

> *Sample question:* What is the average shoe size of all 6th grade students in our school?
> *Sample student response:* "I predict that the average shoe size of 6th grade students in our school is size 7. I am guessing that the data would have a normal shape and would all be clustered around the size 7. I also do not think the data will be very widely dispersed because most 6th grade students have a similar-size shoe."

Other possible questions:

- How much time do middle school students spend on the computer each day?
- How much time do middle school students spend in physical activity each week?
- Does the average 6th grade girl's shoe size differ from the average 6th grade boy's shoe size?

4. Ask student pairs to choose one of the questions that they generated the previous day when completing the Discussion Protocol for Statistical Questions handout and predict, in writing, how the center, spread, and shape of actual data might appear. These descriptions should be shared with other student pairs or with the whole class, as time allows.

Close the Lesson

Return to the essential question and learning objective, and ask students to share with a partner what they have learned about describing data. Present the following models to help students summarize their learning, and ask them to capture these summaries in their math notebooks.

VOCABULARY TERM **center**	DEFINITION (IN YOUR OWN WORDS)
HOW DOES THIS TERM DESCRIBE DATA?	EXAMPLES OF MEASURES OF CENTER

VOCABULARY TERM **shape**	DEFINITION (IN YOUR OWN WORDS)
HOW DOES THIS TERM DESCRIBE DATA?	EXAMPLES (INCLUDE DRAWINGS)

VOCABULARY TERM **spread**	DEFINITION (IN YOUR OWN WORDS)
HOW DOES THIS TERM DESCRIBE DATA?	EXAMPLES OF MEASURES OF VARIABILITY (DRAWINGS MAY HELP)

Additional Resources for This Lesson

Extension Task: Invite individual students, pairs, or the class as a whole to choose one of the questions generated during this lesson and actually collect data to answer it. This activity addresses Common Core standards 6.SP.B.4 and 6.SP.B.5.

Figure A | **Discussion Protocol for Statistical Questions**

Directions: In this activity, you will work cooperatively to write, justify, and critique statistical questions.

1. With your partner, come up with two statistical questions to gather data on topics that you feel are important. Be sure to keep in mind the definition of a statistical question. Write your questions in the space provided.

2. Write a justification statement that communicates how your question meets the components of a statistical question.

3. Once all questions and justification statements have been written, meet with another student pair to share your work. Take turns reading your questions and justification statements.

4. Allow the other group to critique your work, and then you may respond to defend your reasoning. Make any necessary revisions to your statistical questions, and record the final version in the space provided.

Statistical Question 1:

Justification Statement:

Revised Question:

Statistical Question 2:

Justification Statement:

Revised Question:

Figure B	**Vocabulary Sort Cards**	
Word Cards	**Description Cards**	**Example Cards**
statistics	Advertising companies use statistics to determine how to best sell a product.	The branch of mathematics that deals with the collection, organization, and interpretation of data.
statistical question	How long do customers stand in line at McDonald's?	A question that expects variability (differences) in the data.
mean	The mean (average) distance that 6th grade students live from Oak Hill Middle School is 1.5 miles.	A measure of center that represents the central balance point of a set of data (equal share).
median	According to a class survey, the median price of athletic shoes is $68.	A measure of center that represents the midpoint in an ordered data list.

(continued)

Figure B | Vocabulary Sort Cards *(continued)*

range	The range from a low temperature of 23° to a high temperature of 52° is 29°.	A measure of variability; the difference between the largest and smallest values in a numerical data set.
skewed distribution	Although the data showed that most students live within six blocks of school, there are some students who live much farther away.	A set of data that is not equally spread. This results from having a few scores falling further to one end. This impacts the shape of the data.
histogram		A bar graph that shows how many data values fall into a certain interval. The width of the bar represents the interval, while the height indicates the number of data items, or frequency.
box plot		A graphical display of data using rectangles and lines extending to the lowest and highest values that shows the center and the spread of a data set.

Figure C | **Describing Data Guided Practice Handout**

The Process for Describing Histograms. Follow along as we describe and compare the histograms by center, spread, and shape. Vocabulary words in the box will help you develop your descriptions.

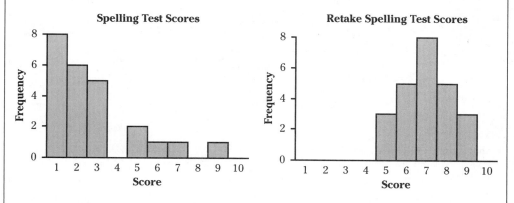

Vocabulary
- *Shape*: normal or symmetrical, skewed or unevenly shaped
- *Center*: lower center, higher center, median
- *Spread*: clustered together, wider range

Use the space below to describe the histograms. Check your statements with a partner once you have finished. Revise the statements as needed.

(continued)

Figure C | **Describing Data Guided Practice Handout** *(continued)*

Describing Box Plots. Follow along as we describe and compare the box plots below by center, spread, and shape. Vocabulary words in the box will help you develop your descriptions.

Recorded Temperature in June for Two Cities

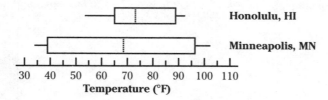

Vocabulary
- *Shape:* skewed right, skewed left, normal
- *Center:* lower center, higher center, median
- *Spread:* clustered together, wider range

Use the space below to describe the box plots. Check your statements with a partner when you have finished. Revise the statements as needed.

Proportional Relationships

Course: 7th grade Mathematics	
Length of Lesson: One hour; one 60-minute class period	

Introduction

In the Common Core State Standards for Mathematics, the Ratios and Proportional Relationships domain is found only in grades 6 and 7. This distinction signifies the unique importance of these concepts and their role as a transition point for learners, who are ready to move beyond the additive relationships found in most elementary arithmetic concepts to multiplicative relationships, which will provide a bridge to Algebra and other higher-level mathematics courses. Although this lesson introduces the concept of proportional relationships, teachers will need to provide multiple experiences with the content in other contexts to cement students' understanding.

Strategies from the Instructional Planning Framework

- *Creating the Environment for Learning:* The essential questions ("What makes a relationship between quantities proportional? How can understanding proportional relationships help us solve daily problems?") and learning objective ("Understand the characteristics of a proportional relationship and use that understanding to solve problems") provide focus for the lesson. The questions and objective should be posted in the classroom and referred to during the lesson. The teacher gives feedback throughout the lesson, and students give feedback to one another through informal partnering. Additionally, a relevant context is used to help students understand how the concepts of this lesson are reflected in situations beyond the mathematics class.

- *Helping Students Develop Understanding:* Nonlinguistic representations are incorporated throughout the lesson as tables, graphs, and equations to represent proportional relationships. Guided practice is used to help students develop an understanding of the concepts, and a student handout provides an organized format for note taking and exploring. Students are also asked to summarize their understanding in lesson-closure activities.

- *Helping Students Extend and Apply Knowledge:* The final task asks students to solve a real-world problem by applying the conceptual understanding and skills they've developed during the lesson.

Common Core State Standards—Knowledge and Skills to Be Addressed

Standards for Mathematical Practice

MP1 Make sense of problems and persevere in solving them.

MP2 Reason abstractly and quantitatively.

MP8 Look for and express regularity in repeated reasoning.

Standards for Mathematical Content

Domain: Ratios and Proportional Relationships

Cluster: Analyze Proportional Relationships and Use Them to Solve Real-World and Mathematical Problems

7.RP.A.2 Recognize and represent proportional relationships between quantities.

a. Decide whether two quantities are in a proportional relationship, e.g., by testing for equivalent ratios in a table or graphing on a coordinate plane and observing whether the graph is a straight line through the origin.

b. Identify the constant of proportionality (unit rate) in tables, graphs, equations, diagrams, and verbal descriptions of proportional relationships.

c. Explain what a point (x, y) on the graph of a proportional relationship means in terms of the situation, with special attention to the points $(0, 0)$ and $(1, r)$ where r is the unit rate.

Common Core State Standards—Prior Knowledge and Skills to Be Applied

Standards for Mathematical Content
Domain: Ratios and Proportional Relationships

Cluster: Analyze Proportional Relationships and Use Them to Solve Real-World Mathematical Problems

7.RP.A.1 Compute unit rates associated with ratios of fractions, including ratios of lengths, areas, and other quantities measured in like or different units.

Cluster: Understand Ratio Concepts and Use Ratio Reasoning to Solve Problems

6.RP.A.1 Understand the concept of a ratio and use ratio language to describe a ratio relationship between two quantities.

6.RP.A.2 Understand the concept of a unit rate a/b associated with a ratio $a:b$ with $b \neq 0$, and use rate language in the context of a ratio relationship.

6.RP.A.3 Use ratio and rate reasoning to solve real-world and mathematical problems, e.g., by reasoning about tables of equivalent ratios, tape diagrams, double number line diagrams, or equations.

Teacher's Lesson Summary

This lesson is designed to help students understand the characteristics of proportional relationships. Students need to be able to distinguish proportional relationships from relationships that aren't proportional and to understand that a proportional relationship is a multiplicative relationship and not an additive relationship. Throughout this lesson, you'll help students demonstrate multiplicative reasoning and incorporate questions that lead students to identify the characteristics of proportional relationships. Students will work in pairs for much of the lesson, and these pairings may be teacher determined or self-selected by students. Whether you set up the pairs or allow the students to do so themselves, the intent of the pair work is for students to help each other focus on the task and check for understanding.

This lesson uses food and beverage preparation to help students understand a practical application of the concept, but it's also important to expose students to other real-world applications of proportional relationship throughout the school year. Model cars, architectural drawings, and picture enlargement/reduction are all potentially helpful examples of proportional relationships found outside a mathematics class.

Essential Questions: What makes a relationship between quantities proportional? How can understanding proportional relationships help us solve daily problems?
Learning Objective: Understand the characteristics of a proportional relationship and use that understanding to solve problems.

Knowledge/Vocabulary Objectives

At the conclusion of this lesson, students will

- Understand the characteristics of a proportional relationship—that it
    - Represents a multiplicative relationship with both quantities increasing by a constant factor.
    - Is represented on a graph as a ray with its endpoint at the origin.
    - Is a collection of pairs of numbers with equal ratios.
- Understand the term *constant of proportionality*.

Skill/Process Objectives

At the conclusion of this lesson, students will be able to

- Identify the constant of proportionality when given tables, graphs, equations, or verbal descriptions.
- Determine whether or not two quantities are in a proportional relationship and justify reasoning.
- Explain what a point (x, y) on the graph of a proportional relationship means in terms of the situation, with special attention to the points $(0, 0)$ and $(1, r)$, where r is the unit rate.

Resources/Preparation Needed

- A handout focused on proportional relationships to structure students' work and keep them engaged during guided instruction (see Figure A, pp. 120–121)

• Various illustrative lesson graphs/tables (see examples throughout this lesson) to project during guided group instruction

Activity Description to Share with Students

Everyone has had the experience of getting together with family and friends to celebrate a special occasion, and food and beverages are often part of the festivities. It takes a solid understanding of proportional reasoning to make sure that there will be enough for everyone and that the food will taste great, whether there are 10 guests or 250 guests. Today's lesson will help you understand proportional relationships and how to apply this understanding to solve everyday problems like scaling up or scaling down a recipe.

Lesson Activity Sequence

Start the Lesson

1. Share the learning objective and essential questions with students, and prompt them to make connections with what they already know about ratios and rates. It might be helpful to get a sense of how many students help out with family cooking duties, as this experience provides context for why exact measurements of recipe ingredients are important for tasty meals. Such discussion also provides formative data about the students' understanding of ratio. Consider presenting the following scenario and questions: *Suppose a fruit punch recipe requires one cup of fruit concentrate to four cups of water. What happens to the taste if you use five cups of water? What happens to the taste if you use three cups of water? Why would keeping a consistent ratio of cups of fruit concentrate to cups of water be important to taste? What does this mean for other recipes?*

Engage Students in Learning the Content

1. Share the following scenario: *A local restaurant uses 6 ounces of diced chicken in one of its entrées. This means that each time this dish is ordered, the restaurant must have 6 ounces of diced chicken available to prepare the dish. How much chicken will the restaurant need if two customers order the dish? If three customers order it? And so on?*

Provide each student with a copy of the **Proportional Relationships Guided Instruction Handout** (**Figure A**, pp. 120–121), which reiterates the scenario and provides students with workspace and a place to take notes. Then proceed with guided instruction. Project a blank grid and ask students to help you set up a graph of the chicken entrée scenario. Make sure that students think about how the x- and y-axes should be labeled. As you create the class grid, students should use the grid on the Proportional Relationships handout and build a graph for their individual use. A completed graph and table are shown here. Be sure to start with the graph before proceeding to the table.

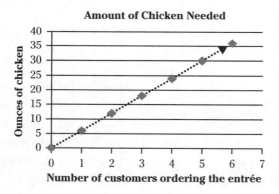

Customers ordering entrée (x)	Ounces of chicken needed (y)
0	0
1	(1 × 6) = 6
2	(2 × 6) = 12
3	(3 × 6) = 18
4	(4 × 6) = 24
5	(5 × 6) = 30
6	(6 × 6) = 36
$y = cx$, or $y = 6x$	

2. Post the following questions, and give student pairs two or three minutes to discuss and respond to them:

   • What are the quantities in this situation, and how are they related?
   • Which quantities vary, and which quantity does not?
   • What does point (3, 18) represent?
   • How would you describe the graph?
   • What is happening at point (0, 0)?
   • What is the unit rate for this situation?
   • How might this information be useful to a restaurant?

3. Reconvene as a whole class, and ask volunteers to offer their responses to the posted questions. Then present the data in table form, using targeted questioning to help students compare the table format to the graph. Your questions might include the following:

- How might the relationship between the x and y variables be stated orally? (*Response:* "For every two entrées ordered, 12 ounces of chicken are needed.")

- What is the unit rate? (*Response:* $\frac{y}{x}$ or $\frac{6}{1}$ or 6.)

- What is an equation that could be used to generalize this relationship? (*Response:* "Total amount of chicken needed = amount of chicken for one serving × the number of customers, or $y = cx$.")

Help students understand that the c in the equation is the unit rate, or *constant of proportionality*. Discuss this vocabulary term with students, and help them understand that it is another term for unit rate.

4. Use the graph of the Amount of Chicken Needed to show how $\frac{y}{x}$ or $\frac{6}{1}$ represents the unit rate or constant of proportionality. Help students interpret this as an increase of six units on the y-axis and an increase of one unit on the x-axis (see the figure below). Ask students to illustrate this $\frac{y}{x}$ relationship on the first graph of the Proportional Relationships handout.

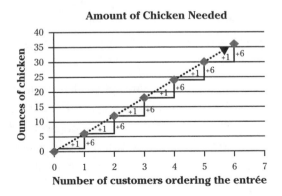

Amount of Chicken Needed

5. Provide students with the plot below and explain to them that it represents two ingredients in a special dipping sauce the restaurant serves with its chicken entrée. For every cup of honey, 1/2 cup of mustard is needed.

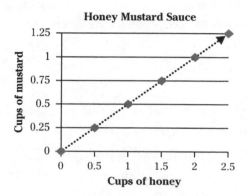

Ask students to work with a partner, and make a list showing how the two plots ("Amount of Chicken Needed" and "Honey Mustard Sauce") are similar. Allow about two minutes, and then bring the class back together. Students' lists of similarities should include statements such as the ones below:

- Each quantity increases (or decreases) by a constant factor.
- Each plot appears to be a ray with its starting point at the origin.
- The points on the graphs are all equal ratios.

Make sure that students understand that these are characteristics of proportional relationships.

6. Ask student pairs to use their understanding of proportional relationships to answer all three parts of question 5 on the Proportional Relationships handout (p. 122). Students should work with a partner and record their responses on the handout. Students may refer to the table to help them think through their responses. *(Answers:* 5b. The constant of proportionality is 1/2; 5c. *y = cx*, or $y = \frac{1}{2}x.$)

7. Because students can develop misconceptions about proportional relationships, it is important to provide non-examples of proportional relationships to help them distinguish between the two. Ask students to indicate whether the examples below are proportional or not proportional, and help them clarify any misconceptions:

a. It takes two people six hours to paint a room. How long does it take four people?

Answer: This isn't a direct proportional relationship because the two quantities don't increase by the same factor. With more people working, the time will decrease rather than increase.

b. If I drive at 60 miles per hour, it will take three hours to reach Johnson City. How long will it take if I drive 70 miles per hour?

Answer: This isn't a direct proportional relationship because the two quantities don't increase by the same factor. An increase in speed will decrease the hours needed to get to Johnson City.

c.

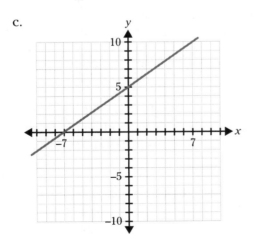

Answer: This graph isn't a proportional relationship as it is not a ray starting from the origin.

d.

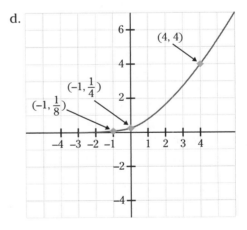

Answer: This graph isn't a proportional relationship as it is not a ray starting from the origin.

e.

x	y
0	10
1	0
2	8.5
3	7.0
8	4.0
9	.5

Answer: This table isn't a proportional relationship since both quantities do not increase by the same factor.

8. Assign the following problem for partner work or independent practice. Walk around the room as students are working to informally assess student understanding, ask probing questions, and identify misconceptions. This will help you to formatively assess and plan for the next lesson. If there isn't enough time to complete the task in class, students may complete it at home or during the next class.

> The student council is hosting a reception to follow an awards program. They want to serve a festive punch and find a recipe that calls for the following:
>
> 1 cup orange juice concentrate
> 2 cups cranberry juice
> 5 cups sparkling water
>
> The council estimates that approximately 55 people will attend the reception, and they expect each person to drink about one cup of punch. Determine how much of each ingredient is needed to serve 55 people. Make sure to show your work and provide the following:
>
> a. A table that shows the proportional relationship of each ingredient
> b. A written statement that describes how much of each ingredient should be purchased and how you know this is sufficient for those attending the reception
> c. The equation that shows the proportional relationship between the orange juice concentrate and the cranberry juice
> d. The equation that shows the proportional relationship between the orange juice concentrate and the sparkling water

When students have completed the problem, ask them to share with a partner (or another student pair) to compare responses and explain their thinking. Then discuss the solution with the entire class.

Close the Lesson

Refer back to the learning objective and essential questions. Ask student pairs to discuss the questions, summarize their learning, and share with the whole group. Next, refer back to the vocabulary term *constant of proportionality*. Ask students to write it in their math notebooks and to provide a description and nonlinguistic representation to help them remember the term.

Figure A | **Proportional Relationships Guided Instruction Handout**

The Chicken Entrée Scenario

A local restaurant uses 6 ounces of diced chicken in one of its entrées. This means that each time this dish is ordered, the restaurant must have 6 ounces of chicken available to prepare the dish. How much chicken will the restaurant need if two customers order the dish? If three customers order it? And so on?

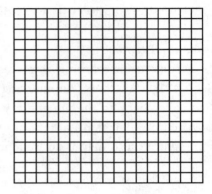

Use the grid to the right to represent the scenario graphically.

Remember:
- Label the axis appropriately.
- Provide a title.
- Plot the points.

Work with a partner to answer the questions below. Be prepared to share your responses with the rest of the class.

1. What are the quantities in this situation, and how are they related?

2. Which quantities vary, and which quantity does not?

3. What does point (3, 18) represent?

4. How would you describe the graph?

5. What is happening at point (0, 0)?

6. What is the unit rate for this situation?

7. How might this information be useful to a restaurant?

Figure A | **Proportional Relationships Guided Instruction Handout** (*continued*)

The table at right represents the same restaurant scenario. Use the data in the table to answer the following questions:

Number of customers ordering entrée (x)	Ounces of chicken needed (y)
0	0
1	6
2	12
3	18
4	24
5	30
6	36

1. How might the relationship between the x and y variables be stated orally?

2. What is the unit rate?

3. What is an equation that could be used to generalize this relationship?

Take a look at the following graph and table that represent the proportions of two ingredients in the dipping sauce that the restaurant serves with its chicken entrée. For every cup of honey, ½ cup of mustard is needed.

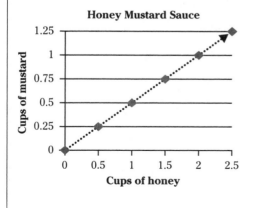

Honey Mustard Sauce

(graph: Cups of mustard vs. Cups of honey)

	Cups of honey (x)	Cups of mustard (y)
No ingredients	0	0
	1	½
Increasing or decreasing ingredient proportions		
½ recipe	½	¼
1 recipe	1	½
	1½	¾
	2	1
	2½	1¼

(*continued*)

Figure A | **Proportional Relationships Guided Instruction Handout** (*continued*)

4. How are the honey mustard sauce plot and the plot for the chicken entrée similar?

5. As written, this recipe serves eight people. What if you only needed to serve four people? Use the graph and chart to answer the following questions:

a. How could proportional reasoning help you determine how much of each ingredient is needed?

b. Because the mustard and honey are in a proportional relationship, what is the constant of proportionality?

c. What is an equation that generalizes this relationship?

Transformations and Congruent Figures

Course: 8th grade Mathematics
Length of Lesson: One hour; one 60-minute class period

Introduction

Even young children realize that objects can move without changing their size and shape. A block might slide across the table, or a puzzle piece may need rotating to fit into place. This intuitive understanding of how shapes move in the plane is developed and formalized through the study of geometric transformations. The Common Core State Standards for Mathematics calls for the formal study of transformations to begin in grade 8. This lesson will provide learning experiences designed to help students understand how transformations, the coordinate plane, and congruency are connected.

Strategies from the Instructional Planning Framework

- *Creating the Environment for Learning:* The essential question ("How can movement of a shape in the plane be analyzed and described?") and learning objective ("To develop an understanding of how transformations affect a shape's orientation, location, size, and shape") provide the focus of the lesson. Clarifying the expected learning outcome helps students understand what they must know, understand, and be able to do to meet the objective, and the feedback they receive gives them a sense of how they are progressing toward that objective. This feedback comes from the teacher, peers, and self-reflection as they participate in the lesson. Cooperative learning takes place both in formal groups, when completing the transformation activity cards, and through informal partnering, as students make connections between transformations and congruent figures.

- *Helping Students Develop Understanding:* Students are prompted to call upon their prior knowledge of transformations to make connections to congruent figures. They use nonlinguistic representations (kinesthetic movements, figures, and grids) to help them develop their problem-solving and spatial reasoning skills. Additionally, the questions in the lesson require students to use analytical thinking and make generalizations.
- *Helping Students Extend and Apply Knowledge:* Students build their own pre-images, apply at least two transformations, and then ask a partner to determine how the image was formed.

Common Core State Standards—Knowledge and Skills to Be Addressed

Standards for Mathematical Content
Domain: Geometry
Cluster: Understand Congruence and Similarity Using Physical Models, Transparencies, or Geometry Software

8.G.A.2 Understand that a two-dimensional figure is congruent to another if the second can be obtained from the first by a sequence of rotations, reflections, and translations; given two congruent figures, describe a sequence that exhibits the congruence between them.

8.G.A.3 Describe the effect of dilations, translations, rotations, and reflections on two-dimensional figures using coordinates.

Common Core State Standards—Prior Knowledge and Skills to Be Applied

Standards for Mathematical Practice
MP4 Model with mathematics.
MP5 Use appropriate tools strategically.
MP7 Look for and make use of structure.

Standards for Mathematical Content
Domain: Geometry

Cluster: Understand Congruence and Similarity Using Physical Models, Transparencies, or Geometry Software

8.G.A.1 Verify experimentally the properties of rotations, reflections, and translations:

a. Lines are taken to lines, and line segments to line segments of the same length.
b. Angles are taken to angles of the same measure.
c. Parallel lines are taken to parallel lines.

Cluster: Solve Real-World and Mathematical Problems Involving Area, Surface Area, and Volume

6.G.A.3 Draw polygons in the coordinate plane given coordinates for the vertices; use coordinates to find the length of a side joining points with the same first coordinate or the same second coordinate. Apply these techniques in the context of solving real-world and mathematical problems.

Teacher's Lesson Summary

In this lesson, students examine transformations of plane figures to describe the figures' movement in the coordinate plane and build a deeper understanding of congruency. In previous grades, students developed an understanding of what it means for two figures to be congruent, but they did not look at this concept through the lens of transformation. Using the coordinate plane, students will examine how coordinate points change during translation, reflection, and rotation, and they will identify the sequence of transformations used to move from a pre-image to the transformed image. Although the lesson can be implemented using basic paper-and-pencil resources, dynamic geometry software can be used to make the exploration of transformation more interactive.

It's important to note the difference between the 8th grade content requirements for this concept and the high school requirements. In 8th grade, students are expected to develop a physical understanding of congruency by using models and demonstrating sequences of transformations. Students observe each figure's properties and determine what sequence of transformations can illustrate the congruence of two figures. At the high school level, students formalize geometric definitions to develop the understanding needed for formal proofs involving congruence and similarity. They also use transformations to build new functions from existing functions, allowing students to identify more optimal models as they analyze complex situations.

Essential Question: How can movement of a shape in the plane be analyzed and described?

Learning Objective: To develop an understanding of how transformations affect a shape's orientation, location, size, and shape.

Knowledge/Vocabulary Objectives

At the conclusion of this lesson, students will

- Understand that shapes can be described in terms of their location in the plane.
- Understand that transformations can be described and quantified using the coordinate system.
- Understand that a two-dimensional figure is congruent to another if the second can be obtained from the first by a sequence of rotations, reflections, and translations.
- Understand and be able to use the terms *congruent, pre-image, transformation, translation, reflection,* and *rotation.*

Skill/Process Objectives

At the conclusion of the lesson, students will be able to

- Describe a sequence of movements (translation, reflection, rotation) that exhibits the congruence between two congruent two-dimensional figures.
- Describe the effect of translations, rotations, and reflections on two-dimensional figures using coordinates.

Resources/Preparation Needed

- Prepared sets of cards for the kinesthetic transformation activity (see Figure A, p. 132), precut and placed in plastic bags for easy distribution, one set for every three-person student group
- Precut 1.5-yard lengths of painter's tape, two lengths for each three-person student group

- Various illustrative images on grid paper or projected from dynamic geometry software (see examples throughout this lesson) to demonstrate various transformations
- A prepared handout to guide students' guided, collaborative, and independent practice using transformations (see Figure B, pp. 133–137), one per student
- Transparency paper and transparency markers for student use
- *Optional:* Dynamic geometry software that students can use during the lesson

Activity Description to Share with Students

Shapes can change location and orientation without changing their size and shape. This lesson will help you develop your problem-solving skills and spatial reasoning skills and build a deeper understanding of congruency as you examine the movement of shapes in the coordinate plane. Outside mathematics class, you'll find these skills and concepts used in the computer animation, art, and construction trades. You'll work in small groups, with a partner, and independently to complete the lesson activities.

Lesson Activity Sequence

Start the Lesson

1. Share the essential question and learning objective with students. Encourage discussion of the question, and help students make connections to previous learning. An out-of-school example might be games such as Bingo and Battleship that involve determining a location on a grid, and an in-school example would be past learning about polygons in the coordinate plane.
2. Provide the vocabulary words for review (*congruent, pre-image, image, transformation, translation, reflection, rotation*), and ask student pairs to discuss strategies that will help them remember these terms. Ask volunteers to share a few of their ideas with the whole group.

Engage Students in Learning the Content

1. Share images, such as the following examples, and ask student pairs to identify and describe the transformations. Encourage students to use vocabulary such as *line of reflection, clockwise or counterclockwise rotation*, and *direction of translation* to add more detail to their descriptions. Clarify any misconceptions as students discuss the image pairs.

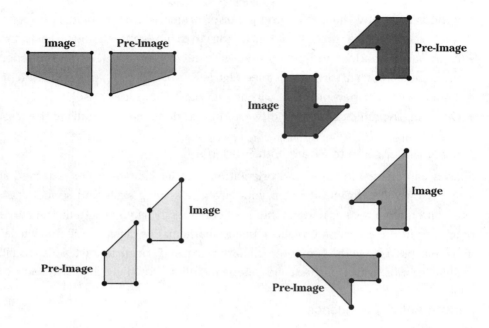

2. Tell students that the coordinate plane can be used to identify a figure's location and describe its movement and that today's lesson will begin with everyone using their bodies to physically represent figures on the coordinate plane. Divide the class into groups of three, and explain to students that they will be working to construct a coordinate grid and transform themselves from the pre-image to the image. Give each group a set of **Kinesthetic Transformation Cards** (see **Figure A**, p. 132), which describe eight different transformations. Move to an open space—in the classroom or at another location (hallway, gym floor, a paved area outdoors)—and explain the activity in greater detail:

Working in groups of three, perform the transformation on each activity card, and record the location of the transformed image in your notebook.
  1. *Create a grid.* Use tape (painter's tape) to construct the x- and y-axes. Identify units 1 to 4 and –1 to –4 on each axis, leaving enough room for a person to stand on the point.

2. *Set roles.* Decide who in your group will perform each required role for Trans-formation A. One of you will be the reader, one will represent the pre-image, and one will represent the transformed image. For each new transformation, switch roles so that everyone has the opportunity to perform each role.

3. *Perform and record the transformation.* Each student should set up a page in his or her notebook to record the location of the pre-image and the trans-formed image. The reader will then draw one card from the bag and read the directions. The "pre-image" will stand on the given location on the coor-dinate grid (pre-image), and the "image" will move as the card describes to the transformed image's location. Each team member will then record the coordinate point for the image in his or her mathematics notebook.

Move about the room (or other area) as student groups perform the transforma-tions, asking questions about the locations and movements to acquire formative information about students' knowledge of transformations on the coordinate grid. When groups have completed the activity cards, bring the whole class back together to check the image's location points and discuss the activity. Be sure to clarify any misconceptions.

3. Share with students that they will next investigate transformations of two-dimensional figures on the coordinate plane. Provide the following (or other) examples and discuss the transformations as a guided group activity. (Dynamic geometry software or grid paper and transparencies can be used for these dem-onstrations.) The goal of this activity is to ensure students can do the following:

- Identify the coordinate points for each figure's vertices.
- Distinguish labels for the pre-image, noted as plain numbers (A, B, C, etc.), and labels for the image, noted with a prime symbol (A′, B′, C′, etc.).
- Describe how the coordinate points change as the pre-image is reflected, translated, and rotated.
- Explain how they know the two figures are congruent (length of segments, angle measurement). It is important to help students understand that reflec-tion, rotation, and translation involve rigid movements. These transforma-tions may change orientation and location, but they do not affect the size or shape of the figures.

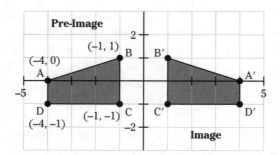

Reflection over the *y*-axis

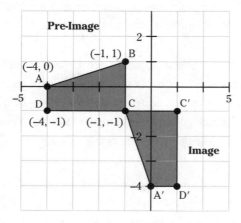

90° counterclockwise rotation about the origin

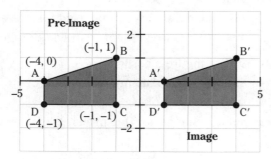

Translation five units along the *x*-axis in the positive direction

4. After students' responses demonstrate they understand how transformations affect figures in the coordinate plane, tell them that they will use their problem-solving skills to determine which transformations were applied to each pre-image in order to form the transformed image. Distribute the **Transformations Handout** (see **Figure B,** pp. 133–137), transparency paper, and a transparency marker to each student. Guide the students through problems 1 and 2, modeling responses, and then ask students to complete the handout, working in pairs or independently, as each item directs. They will use the transparency paper to trace the pre-image, noting the coordinate points as well as the labels for the vertices. Then they will determine a sequence of two transformations that can move the pre-image from its original location to that of the transformed image and record these transformations in the space provided, using appropriate terminology. (*Modeled response:* "The first transformation is a translation. The pre-image moved four units along the *x*-axis in the positive direction and then two units along the *y*-axis in the positive direction.") Next, students will label the vertices and coordinate points for the new image. Encourage students to look for other ways to transform the pre-image using two transformations. Move about the room as students are working to elicit discussion about the concepts. Depending on the time available, problem 7 (pair-sharing of student-created pre-images) can be completed for homework and shared with a partner during the next class period.

Close the Lesson

For the final 10 minutes of the class period, bring the students' focus back to the essential question and the learning objective. Discuss the responses to handout questions 9–11, and ask students to complete a quick-write (two to three sentences) explaining how rigid transformations and congruency are connected. Collect this quick-write and use it as formative data to help assess student understanding of rigid transformations and congruency.

Additional Resources for This Lesson

Online Instructional Unit
National Council of Teachers of Mathematics: "Illuminations—Understanding Congruence, Similarity, and Symmetry Using Transformations and Interactive Figures." Available: http://illuminations.nctm.org/LessonDetail.aspx?ID=U134

Figure A | **Kinesthetic Transformation Cards**

Transformation A

The pre-image is located at point (3, 1). Reflect the pre-image across the *x*-axis. At what point is the image located? Record the response in your notebook.

Transformation B

The pre-image is located at point (2, –2). Translate the pre-image three units in the negative direction along the *x*-axis and one unit in the positive direction along the *y*-axis. At what point is the image located? Record the response in your notebook.

Transformation C

The pre-image is located at point (–3, 2). Rotate the pre-image 90° in a clockwise motion about the origin. At what point is the image located? Record the response in your notebook.

Transformation D

The pre-image is located at point (–2, –3). Reflect the pre-image across the *y*-axis. At what point is the image located? Record the response in your notebook.

Transformation E

The pre-image is located at point (–4, 1). Translate the pre-image three units in the positive direction along the *x*-axis and two units in the negative direction along the *y*-axis. At what point is the image located? Record the response in your notebook.

Transformation F

The pre-image is located at point (3, –1). Rotate the pre-image 180° in a counterclockwise motion about the origin. At what point is the image located? Record the response in your notebook.

Transformation G

The pre-image is located at point (0, 3). Rotate the pre-image 90° in a counterclockwise motion about the origin. At what point is the image located? Record the response in your notebook.

Transformation H

The pre-image is located at point (0, –2). Translate the image two units in the positive direction along the *x*-axis and one unit in the negative direction along the *y*-axis. At what point is the image located? Record the response in your notebook.

Solutions: A. point (3, –1); B. point (–1, –1); C. point (2, 3); D. point (2, –3); E. point (–1, –1); F. point (–3, 1); G. point (–3, 0); H. point (2, –3)

Figure B | **Transformations Handout**

Directions: For problems 1–6, determine the series of transformations (reflection, rotation, and/or translation) applied to each pre-image to form the image. Trace the pre-image on transparency paper as you explore the transformations. Be sure to

   a. Identify coordinate points for both the pre-image and image.
   b. Describe the movements (transformation) of the pre-image to the image.
   c. Explain how you know that each image is congruent to the pre-image.
   d. Determine if there is more than one way to move from the pre-image to the image. If so, describe it.

1. Complete with teacher guidance (guided practice).

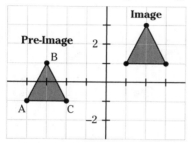

2. Complete with teacher guidance (guided practice).

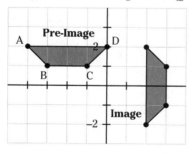

3. Work with a partner.

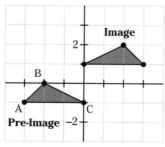

(continued)

Figure B | **Transformations Handout** *(continued)*

4. Work with a partner.

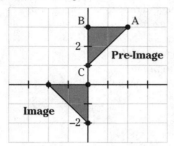

5. Complete independently.

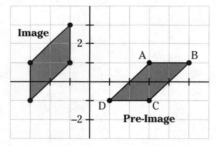

6. Complete independently.

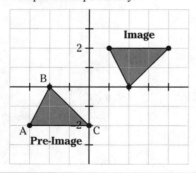

Figure B | **Transformations Handout** *(continued)*

7. *Pair-share.* Draw a figure (pre-image) and perform at least two transformations on the pre-image to form an image. Exchange papers with a partner and ask your partner to determine which transformations were applied. Make sure to identify the coordinate points for the vertices and label the figures appropriately.

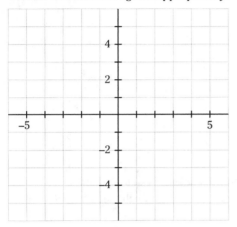

Work independently. Use your understanding of transformations and the coordinate plane to answer the following questions:

8. What happens to the x and y coordinates when an image is reflected across the y-axis? Refer to the figure below to help explain your thinking.

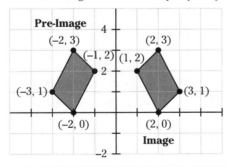

(continued)

Figure B | **Transformations Handout** *(continued)*

9. What happens to the *x* and *y* coordinates when an image is reflected across the *x*-axis? Refer to the figure below to help explain your thinking.

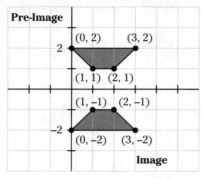

10. Examine the movement of the figure as it is rotated 90°, 180°, and 270° counterclockwise. What point is the center of rotation? How do you know? As you look at the four figures together, what picture comes to mind? What careers might need knowledge of transformation to help in their work?

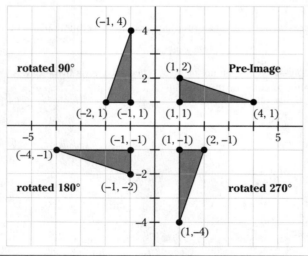

Figure B | **Transformations Handout** *(continued)*

11. Describe how the translated figure moves along the *x*- and *y*-axes.

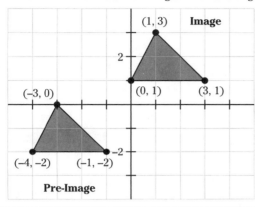

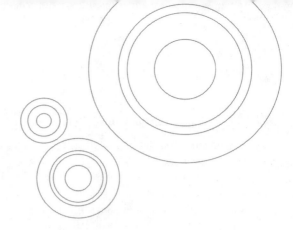

References

Common Core State Standards Initiative. (2010a). *Application of Common Core State Standards for English language learners.* Washington, DC: CCSSO & National Governors Association. Retrieved from http://www.corestandards.org/assets/application-for-english-learners.pdf

Common Core State Standards Initiative. (2010b). *Application to students with disabilities.* Washington, DC: CCSSO & National Governors Association. Retrieved from http://www.corestandards.org/assets/application-to-students-with-disabilities.pdf

Common Core State Standards Initiative. (2010c). *Common Core State Standards for mathematics.* Washington, DC: CCSSO & National Governors Association. Retrieved from http://www.corestandards.org/assets/CCSSI_Math%20Standards.pdf

Common Core State Standards Initiative. (2010d). *Common Core State Standards for mathematics. Appendix A: Designing high school mathematics courses based on the Common Core State Standards.* Washington, DC: CCSSO & National Governors Association. Retrieved from http://www.corestandards.org/assets/CCSSI_Mathematics_Appendix_A.pdf

Common Core Standards Writing Team. (2011, April 17). *Progressions for the Common Core State Standards in mathematics* (draft). Available: http://commoncoretools.files.wordpress.com/2011/04/ccss_progression_nbt_2011_04_073.pdf

Dean, C. B., Hubbell, E. R., Pitler, H., & Stone, B. (2012). *Classroom instruction that works: Research-based strategies for increasing student achievement* (2nd ed.). Alexandria, VA: ASCD.

Kendall, J. S. (2011). *Understanding Common Core State Standards.* Alexandria, VA: ASCD.

National Council of Teachers of Mathematics. (2000). *Principles and standards for school mathematics.* Reston, VA: Author.

National Council of Teachers of Mathematics. (2012). *E-examples from principles and standards for school mathematics.* Available: http://www.nctm.org/standards/content.aspx?id=24600

National Research Council. (2001). *Adding it up: Helping children learn mathematics.* Washington, DC: National Academies Press.

Partnership for Assessment of Readiness for College and Careers. (2011, October). *PARCC model content frameworks: Mathematics grades 3–8 only.* Retrieved from http://www.parcconline.org/sites/parcc/files/PARCCMCFfor3-8Mathematics Fall2011Release.pdf

Schmidt, W. (2012). *Common Core State Standards math: The relationship between high standards, systemic implementation and student achievement.* Lansing, MI: Michigan State University.

Schoenfeld, A., Burkhardt, H., Abedi, J., Hess, K., & Thurlow, M. (2012, March). *Content specifications for the summative assessment of the Common Core State Standards for mathematics* (draft). Olympia, WA: Smarter Balanced Consortium. Available: http://www.smarterbalanced.org/wordpress/wp-content/uploads/2011/12/Math-Content-Specifications.pdf

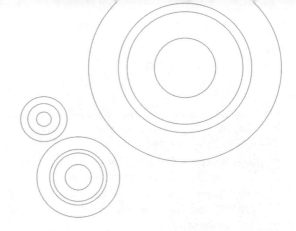

About the Authors

Amitra Schwols serves as a consultant at Mid-continent Research for Education and Learning (McREL). As an analyst at McREL, she has reviewed, revised, and developed standards documents for many districts, state agencies, and organizations. She has also reviewed instructional materials, created lesson plans, and conducted research on a wide variety of education topics. Ms. Schwols's work with the Common Core State Standards includes developing gap analysis, crosswalk, and transition documents, as well as facilitating implementation with groups of teacher leaders. She was a consulting state content expert for mathematics during the development of the Common Core standards and a state consultant to the Partnership for Assessment of Readiness for College and Careers (PARCC) consortium. A former classroom teacher at the secondary grades and a Navy veteran, Ms. Schwols holds a BS in science with an emphasis in physics and mathematics and a minor in English from Colorado State University.

 **Kathleen Dempsey** is a principal consultant with McREL. In this role, Ms. Dempsey works to provide services, strategies, and materials to support improvement in mathematics education, curriculum development, formative assessment, and integration of instructional technology. Additionally, Ms. Dempsey currently serves as a co-principal investigator for two mathematics studies funded by the Institute of Education Sciences. Ms. Dempsey holds an MEd in educational supervision from the College of William and Mary, a BS in elementary education from Old Dominion University, and endorsements in mathematics—Algebra I, gifted education, and middle school education. Before coming to McREL, Ms. Dempsey served as the secondary mathematics coordinator for Virginia Beach City Public Schools in Virginia Beach, Virginia.

John Kendall (Series Editor) is Senior Director in Research at McREL in Denver. Having joined McREL in 1988, Mr. Kendall conducts research and development activities related to academic standards. He directs a technical assistance unit that provides standards-related services to schools, districts, states, and national and international organizations. He is the author of *Understanding Common Core State Standards,* the senior author of *Content Knowledge: A Compendium of Standards and Benchmarks for K–12 Education,* and the author or coauthor of numerous reports and guides related to standards-based systems. These works include *High School Standards and Expectations for College and the Workplace, Essential Knowledge: The Debate over What American Students Should Know,* and *Finding the Time to Learn: A Guide.* He holds an MA in Classics and a BA in English Language and Literature from the University of Colorado at Boulder.

About McREL

McREL is a nationally recognized nonprofit education research and development organization headquartered in Denver, Colorado, with offices in Honolulu, Hawaii, and Omaha, Nebraska. Since 1966, McREL has helped translate research and professional wisdom about what works in education into practical guidance for educators. Our more than 120 staff members and affiliates include respected researchers, experienced consultants, and published writers who provide educators with research-based guidance, consultation, and professional development for improving student outcomes.

ASCD and Common Core State Standards Resources

ASCD believes that for the Common Core State Standards to have maximum effect, they need to be part of a well-rounded whole child approach to education that ensures students are healthy, safe, engaged, supported, and challenged.

For a complete and updated overview of ASCD's resources related to the Common Core standards, including other Quick-Start Guides in the Understanding the Common Core Standards Series, professional development institutes, online courses, links to webinars and to ASCD's free EduCore™ digital tool, and lots more, please visit us at **www.ascd.org/commoncore**.